Wireless Networks and Communications

Wireless Networks and Communications

Samson Colon

WILLFORD PRESS

www.willfordpress.com

Published by Willford Press,
118-35 Queens Blvd., Suite 400,
Forest Hills, NY 11375, USA

ISBN: 978-1-68285-727-4

Cataloging-in-Publication Data

Wireless networks and communications / Samson Colon.
p. cm.
Includes bibliographical references and index.
ISBN 978-1-68285-727-4
1. Wireless communication systems. 2. Wireless sensor networks.
3. Wireless LANs. 4. Multimedia communications.
I. Colon, Samson.
TK5103.2 .W57 2019
621.384--dc23

For information on all Willford Press publications
visit our website at www.willfordpress.com

Contents

Permissions

Index

Preface

Computer networks, which use wireless data connection between network nodes, are called wireless networks. Wireless telecommunication is achieved by radio communication. The different types of wireless networks are wireless personal area networks, wireless local area networks, wireless ad hoc networks, wireless metropolitan area networks, cellular networks, etc. Wireless communication is used in terrestrial microwave, communications satellites, cellular and PCS systems, and free-space optical communication, besides many others. The topics included in this book on wireless networks are of the utmost significance and bound to provide incredible insights to readers. While understanding the long-term perspectives of the topics, it makes an effort in highlighting their impact as a modern tool for the growth of the discipline. In this textbook, constant effort has been made to make the understanding of the difficult concepts of wireless networks as easy and informative as possible, for the readers.

A foreword of all Chapters of the book is provided below:

Chapter 1- A wireless network is the network, which enables wireless data communication between network nodes. This is an introductory chapter, which introduces briefly all the significant aspects of wireless networking such as wireless networking standards, wireless access point, wireless distribution system, wireless application protocol, etc.; **Chapter 2**- There are different categories of wireless networks, depending on the number of devices that a network can connect, or the extent of its functioning. These are Wireless LAN, WAN, PAN, MAN, cellular network, etc., which have been discussed in great detail in this chapter; **Chapter 3**- Wireless sensor networks are a group of sensors that are spatially dispersed and meant for recording and monitoring crucial information of the physical environment, such as temperature, pollution levels, wind, humidity, etc. The aim of this chapter is to explore the central aspects of wireless sensor networks, such as sensor grid and node, location estimation, visual and virtual sensor network, among others; **Chapter 4**- Wireless communication refers to the process of information exchange between two or more remote points that are not connected by an electrical conductor. This transfer of information is achieved by radio waves. All the diverse technologies and devices of wireless communications have been carefully analyzed in this chapter such as mobile telephony, Bluetooth, Wi-Fi and infrared communication.

I would like to thank the entire editorial team who made sincere efforts for this book and my family who supported me in my efforts of working on this book. I take this opportunity to thank all those who have been a guiding force throughout my life.

Samson Colon

Wireless Networking: An Introduction

A wireless network is the network, which enables wireless data communication between network nodes. This is an introductory chapter, which introduces briefly all the significant aspects of wireless networking such as wireless networking standards, wireless access point, wireless distribution system, wireless application protocol, etc.

One of the most transformative technology trends of the past decade is the availability and growing expectation of ubiquitous connectivity. Whether it is for checking email, carrying a voice conversation, web browsing, or myriad other use cases, we now expect to be able to access these online services regardless of location, time, or circumstance: on the run, while standing in line, at the office, on a subway, while in flight, and everywhere in between. Today, we are still often forced to be proactive about finding connectivity (e.g., looking for a nearby WiFi hotspot) but without a doubt, the future is about ubiquitous connectivity where access to the Internet is omnipresent.

Wireless networks are at the epicenter of this trend. At its broadest, a wireless network refers to any network not connected by cables, which is what enables the desired convenience and mobility for the user.

Wireless networks are computer networks that are not connected by cables of any kind. The use of a wireless network enables enterprises to avoid the costly process of introducing cables into buildings or as a connection between different equipment locations. The basis of wireless systems are radio waves, an implementation that takes place at the physical level of network structure.

Given the myriad different use cases and applications, we should also expect to see dozens of different wireless technologies to meet the needs, each with its own performance characteristics and each optimized for a specific task and context. Today, we already have over a dozen widespread wireless technologies in use: WiFi, Bluetooth, ZigBee, NFC, WiMAX, LTE, HSPA, EV-DO, earlier 3G standards, satellite services, and more.

As such, given the diversity, it is not wise to make sweeping generalizations about performance of wireless networks. However, the good news is that most wireless technologies operate on common principles, have common trade-offs, and are subject to common performance criteria and constraints. Once we uncover and understand these fundamental principles of wireless performance, most of the other pieces will begin to automatically fall into place.

Further, while the mechanics of data delivery via radio communication are fundamentally different from the tethered world, the outcome as experienced by the user is, or should be, all the same—same performance, same results. In the long run all applications are and will be delivered over wireless networks; it just may be the case that some will be accessed more frequently over wireless than others. There is no such thing as a *wired application*, and there is zero demand for such a distinction.

All applications should perform well regardless of underlying connectivity. As a user, you should not care about the underlying technology in use, but as developers we must think ahead and architect our applications to anticipate the differences between the different types of networks. And the good news is every optimization that we apply for wireless networks will translate to a better experience in all other contexts.

Wireless Links

Computers are very often connected to networks using wireless links, e.g. WLANs

- *Terrestrial microwave* – Terrestrial microwave communication uses Earth-based transmitters and receivers resembling satellite dishes. Terrestrial microwaves are in the low gigahertz range, which limits all communications to line-of-sight. Relay stations are spaced approximately 48 km (30 mi) apart.

- *Communications satellites* – Satellites communicate via microwave radio waves, which are not deflected by the Earth's atmosphere. The satellites are stationed in space, typically in geosynchronous orbit 35,400 km (22,000 mi) above the equator. These Earth-orbiting systems are capable of receiving and relaying voice, data, and TV signals.

- *Cellular and PCS systems* use several radio communications technologies. The systems divide the region covered into multiple geographic areas. Each area has a low-power transmitter or radio relay antenna device to relay calls from one area to the next area.

- *Radio and spread spectrum technologies* – Wireless local area networks use a high-frequency radio technology similar to digital cellular and a low-frequency radio technology. Wireless LANs use spread spectrum technology to enable communication between multiple devices in a limited area.

- *Free-space optical communication* uses visible or invisible light for communications. In most cases, line-of-sight propagation is used, which limits the physical positioning of communicating devices.

Types of Wireless Networks

Wireless PAN

Wireless personal area networks (WPANs) connect devices within a relatively small area, that is generally within a person's reach. For example, both Bluetooth radio and invisible infrared light provides a WPAN for interconnecting a headset to a laptop. Zig-Bee also supports WPAN applications. Wi-Fi PANs are becoming commonplace (2010) as equipment designers start to integrate Wi-Fi into a variety of consumer electronic devices. Intel "My WiFi" and Windows 7 "virtual Wi-Fi" capabilities have made Wi-Fi PANs simpler and easier to set up and configure.

Wireless LAN

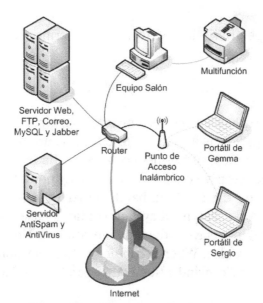

Wireless LANs are often used for connecting to local resources and to the Internet

A wireless local area network (WLAN) links two or more devices over a short distance using a wireless distribution method, usually providing a connection through an access point for internet access. The use of spread-spectrum or OFDM technologies may allow users to move around within a local coverage area, and still remain connected to the network.

Products using the IEEE 802.11 WLAN standards are marketed under the Wi-Fi brand name. Fixed wireless technology implements point-to-point links between computers or networks at two distant locations, often using dedicated microwave or modulated laser light beams over line of sight paths. It is often used in cities to connect networks in two or more buildings without installing a wired link.

Wireless Ad Hoc Network

A wireless ad hoc network, also known as a wireless mesh network or mobile ad hoc network (MANET), is a wireless network made up of radio nodes organized in a mesh topology. Each node forwards messages on behalf of the other nodes and each node performs routing. Ad hoc networks can "self-heal", automatically re-routing around a node that has lost power. Various network layer protocols are needed to realize ad hoc mobile networks, such as Distance Sequenced Distance Vector routing, Associativity-Based Routing, Ad hoc on-demand Distance Vector routing, and Dynamic source routing.

Wireless MAN

Wireless metropolitan area networks are a type of wireless network that connects several wireless LANs.

- WiMAX is a type of Wireless MAN and is described by the IEEE 802.16 standard.

Wireless WAN

Wireless wide area networks are wireless networks that typically cover large areas, such as between neighbouring towns and cities, or city and suburb. These networks can be used to connect branch offices of business or as a public Internet access system. The wireless connections between access points are usually point to point microwave links using parabolic dishes on the 2.4 GHz band, rather than omnidirectional antennas used with smaller networks. A typical system contains base station gateways, access points and wireless bridging relays. Other configurations are mesh systems where each access point acts as a relay also. When combined with renewable energy systems such as photovoltaic solar panels or wind systems they can be stand alone systems.

Cellular Network

A cellular network or mobile network is a radio network distributed over land areas called cells, each served by at least one fixed-location transceiver, known as a cell site or base station. In a cellular network, each cell characteristically uses a different set of radio frequencies from all their immediate neighbouring cells to avoid any interference.

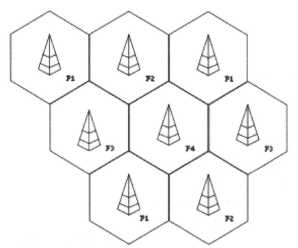

Example of frequency reuse factor or pattern 1/4

When joined together these cells provide radio coverage over a wide geographic area. This enables a large number of portable transceivers (e.g., mobile phones, pagers, etc.) to communicate with each other and with fixed transceivers and telephones anywhere in the network, via base stations, even if some of the transceivers are moving through more than one cell during transmission.

Although originally intended for cell phones, with the development of smartphones, cellular telephone networks routinely carry data in addition to telephone conversations:

- Global System for Mobile Communications (GSM): The GSM network is divided into three major systems: the switching system, the base station system, and the operation and support system. The cell phone connects to the base system station which then connects to the operation and support station; it then connects to the switching station where the call is transferred to where it needs to go. GSM is the most common standard and is used for a majority of cell phones.

- Personal Communications Service (PCS): PCS is a radio band that can be used by mobile phones in North America and South Asia. Sprint happened to be the first service to set up a PCS.

- D-AMPS: Digital Advanced Mobile Phone Service, an upgraded version of AMPS, is being phased out due to advancement in technology. The newer GSM networks are replacing the older system.

Global Area Network

A global area network (GAN) is a network used for supporting mobile across an arbitrary number of wireless LANs, satellite coverage areas, etc. The key challenge in mobile communications is handing off user communications from one local coverage area to the next. In IEEE Project 802, this involves a succession of terrestrial wireless LANs.

Space Network

Space networks are networks used for communication between spacecraft, usually in the vicinity of the Earth. The example of this is NASA's Space Network.

Different uses

Some examples of usage include cellular phones which are part of everyday wireless networks, allowing easy personal communications. Another example, Intercontinental network systems, use radio satellites to communicate across the world. Emergency services such as the police utilize wireless networks to communicate effectively as well. Individuals and businesses use wireless networks to send and share data rapidly, whether it be in a small office building or across the world.

Properties

General

In a general sense, wireless networks offer a vast variety of uses by both business and home users.

> "Now, the industry accepts a handful of different wireless technologies. Each wireless technology is defined by a standard that describes unique functions at both the Physical and the Data Link layers of the OSI model. These standards differ in their specified signaling methods, geographic ranges, and frequency usages, among other things. Such differences can make certain technologies better suited to home networks and others better suited to network larger organizations."

Performance

Each standard varies in geographical range, thus making one standard more ideal than the next depending on what it is one is trying to accomplish with a wireless network. The performance of wireless networks satisfies a variety of applications such as voice and video. The use of this technology also gives room for expansions, such as from 2G to 3G and, 4G and 5G technologies, which stand for the fourth and fifth generation of cell phone mobile communications standards. As wireless networking has become commonplace, sophistication increases through configuration of network hardware and software, and greater capacity to send and receive larger amounts of data, faster, is achieved. Now the wireless network has been running on LTE, which is a 4G mobile communication standard. Users of an LTE network should have data speeds that are 10x faster than a 3G network.

Space

Space is another characteristic of wireless networking. Wireless networks offer many advantages when it comes to difficult-to-wire areas trying to communicate such as across a street or river, a warehouse on the other side of the premises or buildings that are physically separated but operate as one. Wireless networks allow for users to designate a certain space which the network will be able to communicate with other devices through that network.

Space is also created in homes as a result of eliminating clutters of wiring. This technology allows for an alternative to installing physical network mediums such as TPs, coaxes, or fiber-optics, which can also be expensive.

Home

For homeowners, wireless technology is an effective option compared to Ethernet for sharing printers, scanners, and high-speed Internet connections. WLANs help save the cost of installation of cable mediums, save time from physical installation, and also creates mobility for devices connected to the network. Wireless networks are simple and require as few as one single wireless access point connected directly to the Internet via a router.

Wireless Network Elements

The telecommunications network at the physical layer also consists of many interconnected wireline network elements (NEs). These NEs can be stand-alone systems or products that are either supplied by a single manufacturer or are assembled by the service provider (user) or system integrator with parts from several different manufacturers.

Wireless NEs are the products and devices used by a wireless carrier to provide support for the backhaul network as well as a mobile switching center (MSC).

Reliable wireless service depends on the network elements at the physical layer to be protected against all operational environments and applications.

What are especially important are the NEs that are located on the cell tower to the base station (BS) cabinet. The attachment hardware and the positioning of the antenna and associated closures and cables are required to have adequate strength, robustness, corrosion resistance, and resistance against wind, storms, icing, and other weather conditions. Requirements for individual components, such as hardware, cables, connectors, and closures, shall take into consideration the structure to which they are attached.

Difficulties

Interferences

Compared to wired systems, wireless networks are frequently subject to electromagnetic interference. This can be caused by other networks or other types of equipment that generate radio waves that are within, or close, to the radio bands used for communication. Interference can degrade the signal or cause the system to fail.

Absorption and Reflection

Some materials cause absorption of electromagnetic waves, preventing it from reaching the receiver, in other cases, particularly with metallic or conductive materials reflection occurs. This can cause dead zones where no reception is available. Aluminium foiled thermal isolation in modern homes can easily reduce indoor mobile signals by 10 dB frequently leading to complaints about the bad reception of long-distance rural cell signals.

Multipath Fading

In multipath fading two or more different routes taken by the signal, due to reflections, can cause the signal to cancel out at certain locations, and to be stronger in other places (upfade).

Hidden Node Problem

The hidden node problem occurs in some types of network when a node is visible from a wireless access point (AP), but not from other nodes communicating with that AP. This leads to difficulties in media access control.

Shared Resource Problem

The wireless spectrum is a limited resource and shared by all nodes in the range of its transmitters. Bandwidth allocation becomes complex with multiple participating users. Often users are not aware that advertised numbers (e.g., for IEEE 802.11 equipment or LTE networks) are not their capacity, but shared with all other users and thus the individual user rate is far lower. With increasing demand, the capacity crunch is more and more likely to happen. User-in-the-loop (UIL) may be an alternative solution to ever upgrading to newer technologies for over-provisioning.

Capacity

Channel

Shannon's theorem can describe the maximum data rate of any single wireless link, which relates to the bandwidth in hertz and to the noise on the channel.

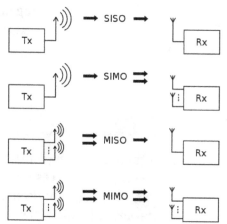

Understanding of SISO, SIMO, MISO and MIMO. Using multiple antennas and transmitting in different frequency channels can reduce fading, and can greatly increase the system capacity.

One can greatly increase channel capacity by using MIMO techniques, where multiple aerials or multiple frequencies can exploit multiple paths to the receiver to achieve much higher throughput – by a factor of the product of the frequency and aerial diversity at each end.

Under Linux, the Central Regulatory Domain Agent (CRDA) controls the setting of channels.

Network

The total network bandwidth depends on how dispersive the medium is (more dispersive medium generally has better total bandwidth because it minimises interference), how many frequencies are available, how noisy those frequencies are, how many aerials are used and whether a directional antenna is in use, whether nodes employ power control and so on. there are two bands for now 2.4 GHz and 5 GHz. mostly 5 gigahertz band gives better connection and speed.

Cellular wireless networks generally have good capacity, due to their use of directional aerials, and their ability to reuse radio channels in non-adjacent cells. Additionally, cells can be made very small using low power transmitters this is used in cities to give network capacity that scales linearly with population density.

Safety

Wireless access points are also often close to humans, but the drop off in power over distance is fast, following the inverse-square law. The position of the United Kingdom's Health Protection Agency (HPA) is that "...radio frequency (RF) exposures from WiFi are likely to be lower than those from mobile phones." It also saw "...no reason why schools and others should not use WiFi equipment." In October 2007, the HPA launched a new "systematic" study into the effects of WiFi networks on behalf of the UK government, in order to calm fears that had appeared in the media in a recent period up to that time". Dr Michael Clark, of the HPA, says published research on mobile phones and masts does not add up to an indictment of WiFi.

Wireless Networking Standards

Home or business owners who are out there searching for their perfect networking gear to implement have to deal with a wide range of choices before coming up to a conclusion. Many of the varieties of technology ranging from 802.11 a/b/n etc. to many other wireless standards available in the market are collectively termed as Wi-Fi technologies. Each of these applications has been built for a specific purpose such as wireless and Bluetooth (other than Wi-Fi).

802.11 and *802.11x* refers to a family of specifications developed by the IEEE for *wireless LAN* (WLAN) technology. 802.11 specifies an over-the-air interface between a wireless client and a base station or between two wireless clients. The IEEE accepted the specification in 1997.

There are several specifications in the *802.11* family:

- 802.11 — applies to wireless LANs and provides 1 or 2 Mbps transmission in the 2.4 GHz band using either frequency hopping spread spectrum (FHSS) or direct sequence spread spectrum (DSSS).

- 802.11a — an extension to 802.11 that applies to wireless LANs and provides up to 54-Mbps in the 5GHz band. 802.11a uses an orthogonal frequency division multiplexing encoding scheme rather than FHSS or DSSS.

- 802.11b (also referred to as 802.11 High Rate or Wi-Fi) — an extension to 802.11 that applies to wireless LANS and provides 11 Mbps transmission (with a fallback to 5.5, 2 and 1-Mbps) in the 2.4 GHz band. 802.11b uses only DSSS. 802.11b was a 1999 ratification to the original 802.11 standard, allowing wireless functionality comparable to Ethernet.

- 802.11e — a wireless draft standard that defines the *Quality of Service* (QoS) support for LANs, and is an enhancement to the 802.11a and 802.11b wireless LAN (WLAN) specifications. 802.11e adds QoS features and multimedia support to the existing IEEE 802.11b and IEEE 802.11a wireless standards, while maintaining full backward compatibility with these standards.

- 802.11g — applies to wireless LANs and is used for transmission over short distances at up to 54-Mbps in the 2.4 GHz bands.

- 802.11n — 802.11n builds upon previous 802.11 standards by adding *multiple-*

input multiple-output (MIMO). The additional transmitter and receiver antennas allow for increased data throughput through spatial multiplexing and increased range by exploiting the spatial diversity through coding schemes like Alamouti coding. The real speed would be 100 Mbit/s (even 250 Mbit/s in PHY level), and so up to 4-5 times faster than 802.11g.

- 802.11ac — 802.11ac builds upon previous 802.11 standards, particularly the 802.11n standard, to deliver data rates of 433Mbps per spatial stream, or 1.3Gbps in a three-antenna (three stream) design. The 802.11ac specification operates only in the 5 GHz frequency range and features support for wider channels (80MHz and 160MHz) and beamforming capabilities by default to help achieve its higher wireless speeds.

- 802.11ac Wave 2 — 802.11ac Wave 2 is an update for the original 802.11ac spec that uses MU-MIMO technology and other advancements to help increase theoretical maximum wireless speeds for the spec to 6.93 Gbps.

- 802.11ad — 802.11ad is a wireless specification under development that will operate in the 60GHz frequency band and offer much higher transfer rates than previous 802.11 specs, with a theoretical maximum transfer rate of up to 7Gbps (Gigabits per second).

- 802.11ah— Also known as Wi-Fi HaLow, 802.11ah is the first Wi-Fi specification to operate in frequency bands below one gigahertz (900 MHz), and it has a range of nearly twice that of other Wi-Fi technologies. It's also able to penetrate walls and other barriers considerably better than previous Wi-Fi standards.

- 802.11r - 802.11r, also called *Fast Basic Service Set* (BSS) Transition, supports VoWi-Fi handoff between access points to enable VoIP roaming on a Wi-Fi network with 802.1X authentication.

- 802.1X — Not to be confused with 802.11x (which is the term used to describe the family of 802.11 standards) 802.1X is an IEEE standard for port-based Network Access Control that allows network administrators to restricted use of IEEE 802 LAN service access points to secure communication between authenticated and authorized devices.

Common Misunderstandings about Achievable Throughput

Across all variations of 802.11, maximum achievable throughputs are given either based on measurements under ideal conditions or in the layer-2 data rates. However, this does not apply to typical deployments in which data is being transferred between two endpoints, of which at least one is typically connected to a wired infrastructure and the other endpoint is connected to an infrastructure via a wireless link.

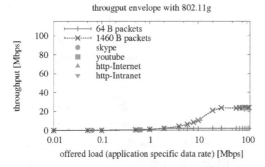

throughput envelope with 802.11g

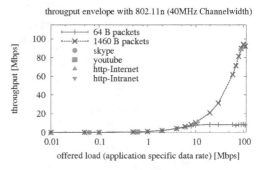

throughput envelope with 802.11n (40MHz Channelwidth)

Graphical representation of Wi-Fi application specific (UDP) performance envelope 2.4 GHz band,
with 802.11g

Graphical representation of Wi-Fi application specific (UDP) performance envelope 2.4 GHz band,
with 802.11n with 40MHz

This means that, typically, data frames pass an 802.11 (WLAN) medium, and are being converted to 802.3 (Ethernet) or vice versa. Due to the difference in the frame (header) lengths of these two media, the application's packet size determines the speed of the data transfer. This means applications that use small packets (e.g., VoIP) create data-flows with high-overhead traffic (i.e., a low goodput). Other factors that contribute to the overall application data rate are the speed with which the application transmits the packets (i.e., the data rate) and, of course, the energy with which the wireless signal is received. The latter is determined by distance and by the configured output power of the communicating devices.

The same references apply to the attached graphs that show measurements of UDP throughput. Each represents an average (UDP) throughput (please note that the error bars are there, but barely visible due to the small variation) of 25 measurements. Each is with a specific packet size (small or large) and with a specific data rate (10 kbit/s – 100 Mbit/s). Markers for traffic profiles of common applications are included as well. These figures assume there are no packet errors, which if occurring will lower transmission rate further.

Channels and Frequencies

802.11b, 802.11g, and 802.11n-2.4 utilize the 2.400–2.500 GHz spectrum, one of the ISM bands. 802.11a, 802.11n and 802.11ac use the more heavily regulated 4.915–5.825 GHz band. These are commonly referred to as the "2.4 GHz and 5 GHz bands" in most sales literature. Each spectrum is sub-divided into *channels* with a center frequency and bandwidth, analogous to the way radio and TV broadcast bands are sub-divided.

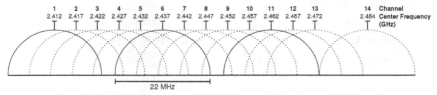

Graphical representation of Wi-Fi channels in the 2.4 GHz band

The 2.4 GHz band is divided into 14 channels spaced 5 MHz apart, beginning with channel 1, which is centered on 2.412 GHz. The latter channels have additional restrictions or are unavailable for use in some regulatory domains.

The channel numbering of the 5.725–5.875 GHz spectrum is less intuitive due to the differences in regulations between countries.

Channel Spacing within the 2.4 GHz band

In addition to specifying the channel center frequency, 802.11 also specifies (in Clause 17) a spectral mask defining the permitted power distribution across each channel. The mask requires the signal be attenuated a minimum of 20 dB from its peak amplitude at ±11 MHz from the centre frequency, the point at which a channel is effectively 22 MHz wide. One consequence is that stations can use only every fourth or fifth channel without overlap.

Availability of channels is regulated by country, constrained in part by how each country allocates radio spectrum to various services. At one extreme, Japan permits the use of all 14 channels for 802.11b, and 1–13 for 802.11g/n-2.4. Other countries such as Spain initially allowed only channels 10 and 11, and France allowed only 10, 11, 12, and 13; however, they now allow channels 1 through 13. North America and some Central and South American countries allow only 1 through 11.

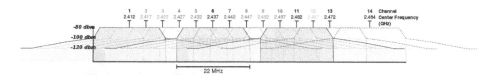

Spectral masks for 802.11g channels 1–14 in the 2.4 GHz band

Since the spectral mask defines only power output restrictions up to ±11 MHz from the center frequency to be attenuated by −50 dBr, it is often assumed that the energy of the channel extends no further than these limits. It is more correct to say that, given the separation between channels, the overlapping signal on any channel should be sufficiently attenuated to minimally interfere with a transmitter on any other channel. Due to the near-far problem a transmitter can impact (desense) a receiver on a "non-overlapping" channel, but only if it is close to the victim receiver (within a meter) or operating above allowed power levels.

Confusion often arises over the amount of channel separation required between transmitting devices. 802.11b was based on direct-sequence spread spectrum (DSSS) modulation and utilized a channel bandwidth of 22 MHz, resulting in *three* "non-overlapping" channels (1, 6, and 11). 802.11g was based on OFDM modulation and utilized a channel bandwidth of 20 MHz. This occasionally leads to the belief that *four* "non-overlapping" channels (1, 5, 9, and 13) exist under 802.11g, although this is not the case as per 17.4.6.3 Channel Numbering of operating channels of the IEEE Std 802.11 (2012), which states "In a multiple cell network topology, overlapping and/or adjacent cells using different channels can operate simultaneously without interference if the distance between the center frequencies is at least 25 MHz." and section 18.3.9.3 and Figure 18-13.

This does not mean that the technical overlap of the channels recommends the non-use of overlapping channels. The amount of interference seen on a configuration using channels 1, 5, 9, and 13 can have very small difference from a three-channel configuration, and in the paper entitled "Effect of adjacent-channel interference in IEEE 802.11 WLANs" by Villegas this is also demonstrated.

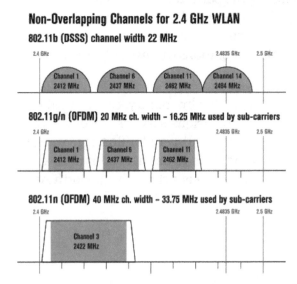

802.11 non-overlapping channels for 2.4GHz. Covers 802.11b,g,n

Although the statement that channels 1, 5, 9, and 13 are "non-overlapping" is limited to spacing or product density, the concept has some merit in limited circumstances. Special care must be taken to adequately space AP cells, since overlap between the channels may cause unacceptable degradation of signal quality and throughput. If more advanced equipment such as spectral analyzers are available, overlapping channels may be used under certain circumstances. This way, more channels are available.

Regulatory Domains and Legal Compliance

IEEE uses the phrase *regdomain* to refer to a legal regulatory region. Different countries define different levels of allowable transmitter power, time that a channel can be occupied, and different available channels. Domain codes are specified for the United States, Canada, ETSI (Europe), Spain, France, Japan, and China.

Most Wi-Fi certified devices default to *regdomain* 0, which means least common denominator settings, i.e., the device will not transmit at a power above the allowable power in any nation, nor will it use frequencies that are not permitted in any nation.

The *regdomain* setting is often made difficult or impossible to change so that the end users do not conflict with local regulatory agencies such as the United States' Federal Communications Commission.

Layer 2 – Datagrams

The datagrams are called *frames*. Current 802.11 standards specify frame types for use in transmission of data as well as management and control of wireless links.

Frames are divided into very specific and standardized sections. Each frame consists of a MAC header, payload, and frame check sequence (FCS). Some frames may not have a payload.

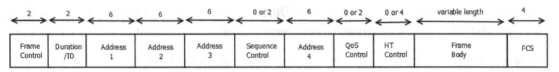

The first two bytes of the MAC header form a frame control field specifying the form and function of the frame. This frame control field is subdivided into the following sub-fields:

- Protocol Version: Two bits representing the protocol version. Currently used protocol version is zero. Other values are reserved for future use.

- Type: Two bits identifying the type of WLAN frame. Control, Data, and Management are various frame types defined in IEEE 802.11.

- Subtype: Four bits providing additional discrimination between frames. Type and Subtype are used together to identify the exact frame.

- ToDS and FromDS: Each is one bit in size. They indicate whether a data frame is headed for a distribution system. Control and management frames set these values to zero. All the data frames will have one of these bits set. However communication within an independent basic service set (IBSS) network always set these bits to zero.

- More Fragments: The More Fragments bit is set when a packet is divided into multiple frames for transmission. Every frame except the last frame of a packet will have this bit set.

- Retry: Sometimes frames require retransmission, and for this there is a Retry bit that is set to one when a frame is resent. This aids in the elimination of duplicate frames.

- Power Management: This bit indicates the power management state of the sender after the completion of a frame exchange. Access points are required to manage the connection, and will never set the power-saver bit.

- More Data: The More Data bit is used to buffer frames received in a distributed system. The access point uses this bit to facilitate stations in power-saver mode. It indicates that at least one frame is available, and addresses all stations connected.

- Protected Frame: The Protected Frame bit is set to one if the frame body is encrypted by a protection mechanism such as Wired Equivalent Privacy (WEP), Wi-Fi Protected Access (WPA), or Wi-Fi Protected Access II (WPA2).

- Order: This bit is set only when the "strict ordering" delivery method is employed. Frames and fragments are not always sent in order as it causes a transmission performance penalty.

The next two bytes are reserved for the Duration ID field. This field can take one of three forms: Duration, Contention-Free Period (CFP), and Association ID (AID).

An 802.11 frame can have up to four address fields. Each field can carry a MAC address. Address 1 is the receiver, Address 2 is the transmitter, Address 3 is used for filtering purposes by the receiver. Address 4 is only present in data frames transmitted between access points in an Extended Service Set or between intermediate nodes in a mesh network.

The remaining fields of the header are:

- The Sequence Control field is a two-byte section used for identifying message order as well as eliminating duplicate frames. The first 4 bits are used for the fragmentation number, and the last 12 bits are the sequence number.

- An optional two-byte Quality of Service control field, present in QoS Data frames; it was added with 802.11e.

The payload or frame body field is variable in size, from 0 to 2304 bytes plus any overhead from security encapsulation, and contains information from higher layers.

The Frame Check Sequence (FCS) is the last four bytes in the standard 802.11 frame. Often referred to as the Cyclic Redundancy Check (CRC), it allows for integrity check of retrieved frames. As frames are about to be sent, the FCS is calculated and appended. When a station receives a frame, it can calculate the FCS of the frame and compare it to the one received. If they match, it is assumed that the frame was not distorted during transmission.

Management Frames

Management frames are not always authenticated, and allow for the maintenance, or discontinuance, of communication. Some common 802.11 subtypes include:

- Authentication frame: 802.11 authentication begins with the wireless network interface card (WNIC) sending an authentication frame to the access point containing its identity. With an open system authentication, the WNIC sends only a single authentication frame, and the access point responds with an authentication frame of its own indicating acceptance or rejection. With shared key authentication, after the WNIC sends its initial authentication request it will receive an authentication frame from the access point containing challenge text. The WNIC sends an authentication frame containing the encrypted version of the challenge text to the access point. The access point ensures the text was encrypted with the correct key by decrypting it with its own key. The result of this process determines the WNIC's authentication status.

- Association request frame: Sent from a station it enables the access point to allocate resources and synchronize. The frame carries information about the WNIC, including supported data rates and the SSID of the network the station wishes to associate with. If the request is accepted, the access point reserves memory and establishes an association ID for the WNIC.

- Association response frame: Sent from an access point to a station containing the acceptance or rejection to an association request. If it is an acceptance, the frame will contain information such an association ID and supported data rates.

- Beacon frame: Sent periodically from an access point to announce its presence and provide the SSID, and other parameters for WNICs within range.

- Deauthentication frame: Sent from a station wishing to terminate connection from another station.

- Disassociation frame: Sent from a station wishing to terminate connection. It's an elegant way to allow the access point to relinquish memory allocation and remove the WNIC from the association table.

- Probe request frame: Sent from a station when it requires information from another station.

- Probe response frame: Sent from an access point containing capability information, supported data rates, etc., after receiving a probe request frame.

- Reassociation request frame: A WNIC sends a reassociation request when it drops from range of the currently associated access point and finds another access point with a stronger signal. The new access point coordinates the forwarding of any information that may still be contained in the buffer of the previous access point.

- Reassociation response frame: Sent from an access point containing the acceptance or rejection to a WNIC reassociation request frame. The frame includes information required for association such as the association ID and supported data rates.

The body of a management frame consists of frame-subtype-dependent fixed fields followed by a sequence of information elements (IEs).

The common structure of an IE is as follows:

```
    ← 1 →  ← 1 →  ← 1-252  →
    ---------------------------------
    | Type |Length|   Data   |
    ---------------------------------
```

Control Frames

Control frames facilitate in the exchange of data frames between stations. Some common 802.11 control frames include:

- Acknowledgement (ACK) frame: After receiving a data frame, the receiving station will send an ACK frame to the sending station if no errors are found. If the sending station doesn't receive an ACK frame within a predetermined period of time, the sending station will resend the frame.

- Request to Send (RTS) frame: The RTS and CTS frames provide an optional collision reduction scheme for access points with hidden stations. A station sends a RTS frame as the first step in a two-way handshake required before sending data frames.

- Clear to Send (CTS) frame: A station responds to an RTS frame with a CTS frame. It provides clearance for the requesting station to send a data frame. The CTS provides collision control management by including a time value for which all other stations are to hold off transmission while the requesting station transmits.

Data Frames

Data frames carry packets from web pages, files, etc. within the body. The body begins with an IEEE 802.2 header, with the Destination Service Access Point (DSAP) specifying the protocol, followed by a Subnetwork Access Protocol (SNAP) header if the DSAP is hex AA, with the organizationally unique identifier (OUI) and protocol ID (PID) fields specifying the protocol. If the OUI is all zeroes, the protocol ID field is an EtherType value. Almost all 802.11 data frames use 802.2 and SNAP headers, and most use an OUI of 00:00:00 and an EtherType value.

Similar to TCP congestion control on the internet, frame loss is built into the operation of 802.11. To select the correct transmission speed or Modulation and Coding Scheme, a rate control algorithm may test different speeds. The actual packet loss rate of an Access points vary widely for different link conditions. There are variations in the loss rate experienced on production Access points, between 10% and 80%, with 30% being a common average. It is important to be aware that the link layer should recover these lost frames. If the sender does not receive an Acknowledgement (ACK) frame, then it will be resent.

Standards and Amendments

Within the IEEE 802.11 Working Group, the following IEEE Standards Association Standard and Amendments exist:

- IEEE 802.11-1997: The WLAN standard was originally 1 Mbit/s and 2 Mbit/s, 2.4 GHz RF and infrared (IR) standard (1997), all the others listed below are Amendments to this standard, except for Recommended Practices 802.11F and 802.11T.

- IEEE 802.11a: 54 Mbit/s, 5 GHz standard (1999, shipping products in 2001)

- IEEE 802.11b: Enhancements to 802.11 to support 5.5 Mbit/s and 11 Mbit/s (1999)

- IEEE 802.11c: Bridge operation procedures; included in the IEEE 802.1D standard (2001)

- IEEE 802.11d: International (country-to-country) roaming extensions (2001)

- IEEE 802.11e: Enhancements: QoS, including packet bursting (2005)

- IEEE 802.11F: Inter-Access Point Protocol (2003) Withdrawn February 2006

- IEEE 802.11g: 54 Mbit/s, 2.4 GHz standard (backwards compatible with b) (2003)

- IEEE 802.11h: Spectrum Managed 802.11a (5 GHz) for European compatibility (2004)

- IEEE 802.11i: Enhanced security (2004)

- IEEE 802.11j: Extensions for Japan (2004)

- IEEE 802.11-2007: A new release of the standard that includes amendments a, b, d, e, g, h, i, and j. (July 2007)

- IEEE 802.11k: Radio resource measurement enhancements (2008)

- IEEE 802.11n: Higher-throughput improvements using MIMO (multiple-input, multiple-output antennas) (September 2009)

- IEEE 802.11p: WAVE—Wireless Access for the Vehicular Environment (such as ambulances and passenger cars) (July 2010)

- IEEE 802.11r: Fast BSS transition (FT) (2008)

- IEEE 802.11s: Mesh Networking, Extended Service Set (ESS) (July 2011)

- IEEE 802.11T: Wireless Performance Prediction (WPP)—test methods and metrics Recommendation _{cancelled}

- IEEE 802.11u: Improvements related to HotSpots and 3rd-party authorization of clients, e.g., cellular network offload (February 2011)

- IEEE 802.11v: Wireless network management (February 2011)

- IEEE 802.11w: Protected Management Frames (September 2009)

- IEEE 802.11y: 3650–3700 MHz Operation in the U.S. (2008)

- IEEE 802.11z: Extensions to Direct Link Setup (DLS) (September 2010)

- IEEE 802.11-2012: A new release of the standard that includes amendments k, n, p, r, s, u, v, w, y, and z (March 2012)

- IEEE 802.11aa: Robust streaming of Audio Video Transport Streams (June 2012)

- IEEE 802.11ac: Very High Throughput <6 GHz; potential improvements over 802.11n: better modulation scheme (expected ~10% throughput increase), wider channels (estimate in future time 80 to 160 MHz), multi user MIMO; (December 2013)

- IEEE 802.11ad: Very High Throughput 60 GHz (December 2012)

- IEEE 802.11ae: Prioritization of Management Frames (March 2012)

- IEEE 802.11af: TV Whitespace (February 2014)

- IEEE 802.11-2016: A new release of the standard that includes amendments ae, aa, ad, ac, and af (December 2016)

- IEEE 802.11ah: Sub-1 GHz license exempt operation (e.g., sensor network, smart metering) (December 2016)

- IEEE 802.11ai: Fast Initial Link Setup (December 2016)

- IEEE 802.11aj: China Millimeter Wave (February 2018)

In Process

- IEEE 802.11ak: General Links *(~ March 2018 for RevCom approval)*

- IEEE 802.11aq: Pre-association Discovery *(~ March 2018 for RevCom approval)*

- IEEE 802.11ax: High Efficiency WLAN *(~ December 2019 for RevCom submittal)*

- IEEE 802.11ay: Enhancements for Ultra High Throughput in and around the 60 GHz Band *(~ November 2019 for final EC approval)*

- IEEE 802.11az: Next Generation Positioning *(~ March 2021 for .11az final)*

- IEEE 802.11ba: Wake Up Radio *(~ July 2020 for RevCom submittal)*

802.11F and 802.11T are recommended practices rather than standards, and are capitalized as such.

802.11m is used for standard maintenance. 802.11ma was completed for 802.11-2007, 802.11mb for 802.11-2012, and 802.11mc for 802.11-2016.

Standard vs. Amendment

Both the terms "standard" and "amendment" are used when referring to the different variants of IEEE standards.

As far as the IEEE Standards Association is concerned, there is only one current standard; it is denoted by IEEE 802.11 followed by the date that it was published. IEEE 802.11-2016 is the only version currently in publication, superseding previous releases. The standard is updated by means of amendments. Amendments are created by task groups (TG). Both the task group and their finished document are denoted by 802.11 followed by a non-capitalized letter, for example, IEEE 802.11a and IEEE 802.11b. Updating 802.11 is the responsibility of task group m. In order to create a new version, TGm combines the previous version of the standard and all published amendments. TGm also provides clarification and interpretation to industry on published documents. New versions of the IEEE 802.11 were published in 1999, 2007, 2012, and 2016.

Nomenclature

Various terms in 802.11 are used to specify aspects of wireless local-area networking operation, and may be unfamiliar to some readers.

For example, Time Unit (usually abbreviated TU) is used to indicate a unit of time equal to 1024 microseconds. Numerous time constants are defined in terms of TU (rather than the nearly equal millisecond).

Also the term "Portal" is used to describe an entity that is similar to an 802.1H bridge. A Portal provides access to the WLAN by non-802.11 LAN STAs.

Community Networks

With the proliferation of cable modems and DSL, there is an ever-increasing market of people who wish to establish small networks in their homes to share their broadband Internet connection.

Many hotspot or free networks frequently allow anyone within range, including passersby outside, to connect to the Internet. There are also efforts by volunteer groups to establish wireless community networks to provide free wireless connectivity to the public.

Security

In 2001, a group from the University of California, Berkeley presented a paper describing weaknesses in the 802.11 Wired Equivalent Privacy (WEP) security mechanism defined in the original standard; they were followed by Fluhrer, Mantin, and Shamir's paper titled "Weaknesses in the Key Scheduling Algorithm of RC4". Not long after, Adam Stubblefield and AT&T publicly announced the first verification of the attack. In the attack, they were able to intercept transmissions and gain unauthorized access to wireless networks.

The IEEE set up a dedicated task group to create a replacement security solution, 802.11i (previously this work was handled as part of a broader 802.11e effort to enhance the MAC layer). The Wi-Fi Alliance announced an interim specification called Wi-Fi Protected Access (WPA) based on a subset of the then current IEEE 802.11i draft. These started to appear in products in mid-2003. IEEE 802.11i (also known as WPA2) itself was ratified in June 2004, and uses the Advanced Encryption Standard AES, instead of RC4, which was used in WEP. The modern recommended encryption for the home/consumer space is WPA2 (AES Pre-Shared Key), and for the enterprise space is WPA2 along with a RADIUS authentication server (or another type of authentication server) and a strong authentication method such as EAP-TLS.

In January 2005, the IEEE set up yet another task group "w" to protect management and broadcast frames, which previously were sent unsecured. Its standard was published in 2009.

In December 2011, a security flaw was revealed that affects some wireless routers with a specific implementation of the optional Wi-Fi Protected Setup (WPS) feature. While WPS is not a part of 802.11, the flaw allows an attacker within the range of the wireless router to recover the WPS PIN and, with it, the router's 802.11i password in a few hours.

In late 2014, Apple announced that its iOS 8 mobile operating system would scramble MAC addresses during the pre-association stage to thwart retail footfall tracking made possible by the regular transmission of uniquely identifiable probe requests.

Non-standard 802.11 Extensions and Equipment

Many companies implement wireless networking equipment with non-IEEE standard 802.11 extensions either by implementing proprietary or draft features. These changes may lead to incompatibilities between these extensions.

Wireless Application Protocol

Wireless application protocol (WAP) is a communications protocol that is used for wireless data access through most mobile wireless networks. WAP enhances wireless specification interoperability and facilitates instant connectivity between interactive wireless devices (such as mobile phones) and the Internet.

WAP functions in an open application environment and may be created on any type of OS. Mobile users prefer WAP because of its ability to efficiently deliver electronic information.

The idea behind WAP is simple: simplify the delivery of Internet content to wireless devices by delivering a comprehensive, Internet-based, wireless specification. The WAP Forum released the first version of WAP in 1998. Since then, it has been widely adopted by wireless phone manufacturers, wireless carriers, and application developers worldwide. Many industry analysts estimate that 90 percent of mobile phones sold over the next few years will be WAP-enabled.

The driving force behind WAP is the WAP Forum component of the Open Mobile Alliance. The WAP Forum was founded in 1997 by Ericsson, Motorola, Nokia, and Openwave Systems (the latter known as Unwired Planet at the time) with the goal of making wireless Internet applications more mainstream by delivering a development specification and framework to accelerate the delivery of wireless applications. Since then, more than 300 corporations have joined the forum, making WAP the de facto standard for wireless Internet applications. In June 2002, the WAP Forum, the Location Interoperability Forum, SyncML Initiative, MMS Interoperability Group, and Wireless Village consolidated under the name Open Mobile Alliance to create a governing body that will be at the center of all mobile application standardization work.

The WAP architecture is composed of various protocols and an XML-based markup language called the Wireless Markup Language (WML), which is the successor to the Handheld Device Markup Language (HDML) as defined by Openwave Systems. WAP 2.x contains a new version of WML, commonly referred to as WML2; it is based on the eXtensible HyperText Markup Language (XHTML), signaling part of WAP's move toward using common Internet specifications such as HTTP and TCP/IP.

Technical Specifications

WAP Protocol Stack

The WAP standard described a protocol suite or stack allowing the interoperability of WAP equipment and software with different network technologies, such as GSM and IS-95 (also known as CDMA).

Wireless Application Environment (WAE)	
Wireless Session Protocol (WSP)	
Wireless Transaction Protocol (WTP)	WAP protocol suite
Wireless Transport Layer Security (WTLS)	
Wireless Datagram Protocol (WDP)	
*** Any wireless data network ***	

The bottom-most protocol in the suite, the Wireless Datagram Protocol (WDP), functions as an adaptation layer that makes every data network look a bit like UDP to the upper layers by providing unreliable transport of data with two 16-bit port numbers (origin and destination). All the upper layers view WDP as one and the same protocol, which has several "technical realizations" on top of other "data bearers" such as SMS, USSD, etc. On native IP bearers such as GPRS, UMTS packet-radio service, or PPP on top of a circuit-switched data connection, WDP is in fact exactly UDP.

WTLS, an optional layer, provides a public-key cryptography-based security mechanism similar to TLS.

WTP provides transaction support (reliable request/response) adapted to the wireless world. WTP supports more effectively than TCP the problem of packet loss, which occurs commonly in 2G wireless technologies in most radio conditions, but is misinterpreted by TCP as network congestion.

Finally, one can think of WSP initially as a compressed version of HTTP.

This protocol suite allows a terminal to transmit requests that have an HTTP or HTTPS equivalent to a WAP gateway; the gateway translates requests into plain HTTP.

The Wireless Application Environment (WAE) space defines application-specific markup languages.

For WAP version 1.X, the primary language of the WAE is Wireless Markup Language (WML). In WAP 2.0, the primary markup language is XHTML Mobile Profile.

WAP Push

WAP Push was incorporated into the specification to allow the WAP content to be pushed to the mobile handset with minimal user intervention. A WAP Push is basically a specially encoded message which includes a link to a WAP address.

WAP Push was specified on top of Wireless Datagram Protocol (WDP); as such, it can be delivered over any WDP-supported bearer, such as GPRS or SMS. Most GSM networks have a wide range of modified processors, but GPRS activation from the network is not generally supported, so WAP Push messages have to be delivered on top of the SMS bearer.

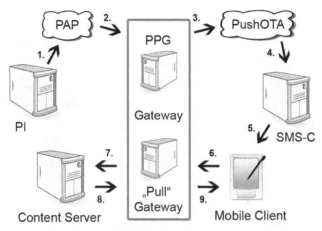

WAP Push process

On receiving a WAP Push, a WAP 1.2 (or later) -enabled handset will automatically give the user the option to access the WAP content. This is also known as WAP Push SI (Service Indication). A variant, known as WAP Push SL (Service Loading), directly opens the browser to display the WAP content, without user interaction. Since this behaviour raises security concerns, some handsets handle WAP Push SL messages in the same way as SI, by providing user interaction.

The network entity that processes WAP Pushes and delivers them over an IP or SMS Bearer is known as a Push Proxy Gateway (PPG).

WAP 2.0

A re-engineered 2.0 version was released in 2002. It uses a cut-down version of XHT-ML with end-to-end HTTP, dropping the gateway and custom protocol suite used to communicate with it. A WAP gateway can be used in conjunction with WAP 2.0; however, in this scenario, it is used as a standard proxy server. The WAP gateway's role would then shift from one of translation to adding additional information to each request. This would be configured by the operator and could include telephone numbers, location, billing information, and handset information.

Mobile devices process XHTML Mobile Profile (XHTML MP), the markup language defined in WAP 2.0. It is a subset of XHTML and a superset of XHTML Basic. A version of Cascading Style Sheets (CSS) called WAP CSS is supported by XHTML MP.

MMS

Multimedia Messaging Service (MMS) is a combination of WAP and SMS allowing for sending of picture messages.

Criticism

Commentators criticized several aspects of Wireless Markup Language (WML) and WAP. Technical criticisms include:

- The idiosyncratic WML language: WML cut users off from the conventional HTML Web, leaving only native WAP content and Web-to-WAP proxi-content available to WAP users. However, others argue that technology at that stage would simply not have been able to give access to anything but custom-designed content which was the sole purpose of WAP and its simple, reduced complexity interface as the citizens of many nations are not connected to the web at the present time and have to use government funded and controlled portals to WAP and similar non-complex services.

- Under-specification of terminal requirements: The early WAP standards included many optional features and under-specified requirements, which meant that compliant devices would not necessarily interoperate properly. This resulted in great variability in the actual behavior of phones, principally because WAP-service implementers and mobile-phone manufacturers did not obtain a copy of the standards or the correct hardware and the standard software modules. As an example, some phone models would not accept a page more than 1 Kb in size; others would downright crash. The user interface of devices was also underspecified: as an example, accesskeys (e.g., the ability to press '4' to access directly the fourth link in a list) were variously implemented depending on phone models (sometimes with the accesskey number automatically displayed by the browser next to the link, sometimes without it, and sometimes accesskeys were not implemented at all).

- Constrained user interface capabilities: Terminals with small black-and-white screens and few buttons, like the early WAP terminals, face difficulties in presenting a lot of information to their user, which compounded the other problems: one would have had to be extra careful in designing the user interface on such a resource-constrained device which was the real concept of WAP.

- Lack of good authoring tools: The problems above might have succumbed in the face of a WML authoring tool that would have allowed content providers to easily publish content that would interoperate flawlessly with many models, adapting the pages presented to the User-Agent type. However, the development kits which existed did not provide such a general capability. Developing for the web was easy: with a text editor and a web browser, anybody could get started, thanks also to the forgiving nature of most desktop browser rendering engines. By contrast, the stringent requirements of the WML specifications, the variability in terminals, and the demands of testing on various wireless terminals, along with the lack of widely available desktop authoring and emulation tools, considerably lengthened the time required to complete most projects. As of 2009, however, with many mobile devices supporting XHTML, and programs such as Adobe Go Live and Dreamweaver offering improved web-authoring tools, it is becoming easier to create content, accessible by many new devices.

- Lack of user agent profiling tools: It quickly became nearly impossible for web hosts to determine if a request came from a mobile device, or from a larger more capable device. No useful profiling or database of device capabilities were built into the specifications in the unauthorized non-compliant products.

Other criticisms address the wireless carriers' particular implementations of WAP:

- Neglect of content providers: Some wireless carriers had assumed a "build it and they will come" strategy, meaning that they would just provide the transport of data as well as the terminals, and then wait for content providers to publish their services on the Internet and make their investment in WAP useful. However, content providers received little help or incentive to go through the complicated route of development. Others, notably in Japan (cf. below), had a more thorough dialogue with their content-provider community, which was then replicated in modern, more successful WAP services such as i-mode in Japan or the Gallery service in France.

- Lack of openness: Many wireless carriers sold their WAP services as "open", in that they allowed users to reach any service expressed in WML and published on the Internet. However, they also made sure that the first page that clients accessed was their own "wireless portal", which they controlled very closely. Some carriers also turned off editing or accessing the address bar in the device's browser. To facilitate users wanting to go off deck, an address bar on a form on a page linked off the hard coded home page was provided. It makes it easier for carriers to implement filtering of off deck WML sites by URLs or to disable the address bar in the future if the carrier decides to switch all users to a walled garden model. Given the difficulty in typing up fully qualified URLs on a phone keyboard, most users would give up going "off portal" or out of the walled garden; by not letting third parties put their own entries on the operators' wireless portal, some contend that operators cut themselves off from a valuable opportunity. On the other hand, some operators argue that their customers would have wanted them to manage the experience and, on such a constrained device, avoid giving access to too many services.

Protocol Design Lessons from WAP

The original WAP model provided a simple platform for access to web-like WML services and e-mail using mobile phones in Europe and the SE Asian regions. In 2009 it continued to have a considerable user base. The later versions of WAP, primarily targeting the United States market, were designed for a different requirement - to enable full web XHTML access using mobile devices with a higher specification and cost, and with a higher degree of software complexity.

Considerable discussion has addressed the question whether the WAP protocol design was appropriate. Some have suggested that the bandwidth-sparing simple interface of Gopher would be a better match for mobile phones and Personal digital assistants (PDAs).

The initial design of WAP specifically aimed at protocol independence across a range of different protocols (SMS, IP over PPP over a circuit switched bearer, IP over GPRS, etc.). This has led to a protocol considerably more complex than an approach directly over IP might have caused.

Most controversial, especially for many from the IP side, was the design of WAP over IP. WAP's transmission layer protocol, WTP, uses its own retransmission mechanisms over UDP to attempt to solve the problem of the inadequacy of TCP over high-packet-loss networks.

WAP Influence on Modern Web Sites

The original WAP model served pages in WML. WML was based on the concept of documents known as "decks", with data in each deck structured into one or more "cards" (pages) – each of which represents a single interaction with the user. With the advent of responsive web design and mobile first approaches the challenges of resizing and adapting content have led to the concept of decks and cards to be recycled. Cards are served in rows and columns to match the device's capabilities, and different decks can be delivered to the device based on the network, device and media support capabilities.

Wireless Access Point

Wireless access points are everywhere these days. In any larger city around the world, it has become impossible to walk a block without stumbling on at least one or two access points.

A wireless access point, commonly called just access point (AP), is a networking device that allows easy access to the Internet over the air. Most access points look very similar to routers. In fact, modern routers can usually function as access points. Internet Service Providers typically give their customers a router with the functionality of an access point to make the set up simpler. If they gave them a router without the access point functionality, the customers would have to connect a dedicated access point to the router to enjoy wireless Internet access, which would be highly inconvenient and beyond the expertise of most home users.

The term wireless access point is often confused with the term hotspot. A wireless access point covers an area with a WiFi signal, and the area in which one can connect to the Internet over the air is called hotspot.

Before WiFi networks, it was quite problematic to connect new devices to the Internet because each new device had to be connected with a wire to an Internet-connected router. Of course, students or office workers seldom carried their own personal electronic devices, so the situation wasn't nearly as bleak as it would be today if we didn't have wireless access points.

After the explosion of the smartphone market, fast, ubiquitous Internet access become the norm and so did WiFi access points. Most people manage their own WiFi access point at home, but not many know how to achieve the best signal strength and the best download and upload speeds possible.

Connections

An AP connects directly to a wired local area network, typically Ethernet, and the AP then provides wireless connections using wireless LAN technology, typically Wi-Fi, for other devices to use that wired connection. APs support the connection of multiple wireless devices through their one wired connection.

Linksys "WAP54G" 802.11g wireless access point

Embedded RouterBoard 112 with U.FL-RSMA pigtail and R52 mini PCI Wi-Fi card widely used by wireless Internet service providers (WISPs) across the world

Wireless Access Point vs. Ad Hoc Network

Some people confuse wireless access points with wireless ad hoc networks. An ad hoc network uses a connection between two or more devices *without* using a wireless access point; The devices communicate directly when in range. Because setup is easy and does not require an access point, an ad hoc network is used in situations such as a quick data exchange or a multiplayer video game. Due to its peer-to-peer layout, ad hoc Wi-Fi connections are similar to connections available using Bluetooth.

Ad hoc connections are generally not recommended for a permanent installation.

Internet access via ad hoc networks, using features like Windows' Internet Connection Sharing, may work well with a small number of devices that are close to each other, but ad hoc networks don't scale well. Internet traffic will converge to the nodes with direct internet connection, potentially congesting these nodes. For internet-enabled nodes, access points have a clear advantage, with the possibility of having a wired LAN.

Limitations

It is generally recommended that one IEEE 802.11 AP should have, at a maximum, 15-25 clients per radio (most APs having between 1 and 4 radios). However, the actual maximum number of clients that can be supported can vary significantly depending on several factors, such as type of APs in use, density of client environment, desired client throughput, etc. The range of communication can also vary significantly, depending on such variables as indoor or outdoor placement, height above ground, nearby obstructions, other electronic devices that might actively interfere with the signal by broadcasting on the same frequency, type of antenna, the current weather, operating radio frequency, and the power output of devices. Network designers can extend the range of APs through the use of repeaters, which amplify a radio signal, and reflectors, which only bounce it. In experimental conditions, wireless networking has operated over distances of several hundred kilometers.

Most jurisdictions have only a limited number of frequencies legally available for use by wireless networks. Usually, adjacent APs will use different frequencies (Channels) to communicate with their clients in order to avoid interference between the two nearby systems. Wireless devices can "listen" for data traffic on other frequencies, and can rapidly switch from one frequency to another to achieve better reception. However, the limited number of frequencies becomes problematic in crowded downtown areas with tall buildings using multiple APs. In such an environment, signal overlap becomes an issue causing interference, which results in signal droppage and data errors.

Wireless networking lags wired networking in terms of increasing bandwidth and throughput. While (as of 2013) high-density 256-QAM (TurboQAM) modulation, 3-antenna wireless devices for the consumer market can reach sustained real-world speeds of some 240 Mbit/s at 13 m behind two standing walls (NLOS) depending on their nature or 360 Mbit/s at 10 m line of sight or 380 Mbit/s at 2 m line of sight (IEEE 802.11ac) or 20 to 25 Mbit/s at 2 m line of sight (IEEE 802.11g), wired hardware of similar cost reaches closer to 1000 Mbit/s up to specified distance of 100 m with twisted-pair cabling in optimal conditions (Category 5 (known as Cat-5) or better cabling with Gigabit Ethernet). One impediment to increasing the speed of wireless communications comes from Wi-Fi's use of a shared communications medium: Thus, two stations in infrastructure mode that are communicating with each other even over the same AP must have each and every frame transmitted twice: from the sender to the AP, then from the AP to the receiver. This approximately halves the effective bandwidth, so an AP is only able to use somewhat less than half the actual over-the-air rate for data throughput. Thus a typical 54 Mbit/s wireless connection actually carries TCP/IP data at 20 to 25 Mbit/s. Users of legacy wired networks expect faster speeds, and people using wireless connections keenly want to see the wireless networks catch up.

By 2012, 802.11n based access points and client devices have already taken a fair share of the marketplace and with the finalization of the 802.11n standard in 2009 inherent problems integrating products from different vendors are less prevalent.

Security

Wireless access has special security considerations. Many wired networks base the security on physical access control, trusting all the users on the local network, but if wireless access points are connected to the network, anybody within range of the AP (which typically extends farther than the intended area) can attach to the network.

The most common solution is wireless traffic encryption. Modern access points come with built-in encryption. The first generation encryption scheme, WEP, proved easy to crack; the second and third generation schemes, WPA and WPA2, are considered secure if a strong enough password or passphrase is used.

Some APs support hotspot style authentication using RADIUS and other authentication servers.

Opinions about wireless network security vary widely. For example, in a 2008 article for *Wired* magazine, Bruce Schneier asserted the net benefits of open Wi-Fi without passwords outweigh the risks, a position supported in 2014 by Peter Eckersley of the Electronic Frontier Foundation.

The opposite position was taken by Nick Mediati in an article for *PC World*, in which he takes the position that every wireless access point should be protected with a password.

WiFi Access Point

As you've most likely experienced numerous times first-hand, the WiFi signal has a limited range and is affected by various obstacles, such as walls, and interference, such as your neighbor's WiFi network.

To optimize your home wireless network, the first step is to discover where the signal is the weakest. Just try to remember all the places in your home where web take forever to lead and where online videos constantly buffer. To get rid of these zones of a weak signal, you need to change the placement of your wireless access point to cover your entire home with an evenly strong WiFi signal.

But moving the WiFi access point randomly could easily make everything much worse. That's where WiFi analysis and visualization tools such as NetSpot come in. With NetSpot, you can quickly create a visual map of your home WiFi network and see where the signal is the strongest and where it is the weakest. You may discover that the half of your home closer to the router is covered with a very strong signal, while the other half leaves a lot to be desired. In that case, the solution is simple: move the access point to the center of your home to achieve even coverage.

In some cases, you may discover that your WiFi access point isn't sufficiently strong to cover all parts of your home no matter where you place it. The solution? Either buy a new access point or install a WiFi booster. Some of the best wireless access points on the market are strong enough to cover even a large apartment or a smaller house, and they are loaded with useful extra features. WiFi boosters, on the other hand, can quickly extend the range of any wireless access point, but they create an additional network which doesn't benefit you unless you manually switch to it. In both cases, NetSpot can help you verify whether you newly purchased access point or a booster has done its job.

Wireless Network Interface Controller

Abbreviated as WNIC, a *wireless network interface card* is a network card which is used to connect radio-based computer networks. WNICs uses an antenna to communicate through microwaves and is typically connected using the computer's PCI bus or USB port. Similar to a Network Interface Card (NIC), the WNIC also works on Layer 1 and Layer 2 of the OSI Model.

WNIC Card Types

Classified by bus interface

- ISA bus network card
- PCI bus network card
- PCI-X bus network card
- PCMCIA bus network card
- USB bus network interface card

ISA card and PCI card is the most common, but the PCI card tends to mainstream status. ISA network card bandwidth is generally 10Mbps, PCI bus network card bandwidth from 10Mbps to 1000Mbps.

Classified by network interface

- RJ-45 interface network card
- BNC interface network card
- AUI interface network card

- FDDI interface network card

- ATM interface network card

The common interfaces are Ethernet RJ-45 interface, thin coaxial cable BNC interface and thick coaxial AUI interface, FDDI interface, ATM interface and so on. And in order to apply to a wider range of applications, some cards providing two or more types of interface, some of them will also provide RJ-45, BNC interface or AUI interface simultaneously.

Classified by bandwidth

- 10Mbps network interface card

- 100Mbps network interface card

- 10Mbps/100Mbps network interface card

- 1000Mbps network interface card

With the development of network technology, network bandwidth is also increasing, but different bandwidth applied to different application, currently the main network card is 10Mbps network card, 100Mbps Ethernet card 10Mbps / 100Mbps adaptive network card, 1000Mbps Gigabit Ethernet card.

Classified by application fields

- Workstation's network interface card

- Server's network interface card

The network interface cards we mentioned above are basically workstation cards, in fact, usually they used in the general server. However, in large networks, the server usually uses a specialized network card. Relative to the workstation used by ordinary card, it has a better performance in bandwidth (usually 100Mbps, the mainstream server network card for the 64-bit Gigabit Ethernet), the number of interfaces, stability, error correction, etc. Some server NIC supports redundant backup, hot-plug and other server-specific functions.

Modes of Operation

An 802.11 WNIC can operate in two modes known as *infrastructure mode* and *ad hoc mode*:

Infrastructure mode

> In an infrastructure mode network the WNIC needs a wireless access point: all data is transferred using the access point as the central hub. All wireless nodes in an infrastructure mode network connect to an access point. All nodes connecting to the

access point must have the same service set identifier (SSID) as the access point, and if a kind of wireless security is enabled on the access point (such as WEP or WPA), they must share the same keys or other authentication parameters.

Ad hoc mode

In an ad hoc mode network the WNIC does not require an access point, but rather can interface with all other wireless nodes directly. All the nodes in an ad hoc network must have the same channel and SSID.

Specifications

The IEEE 802.11 standard sets out low-level specifications for how all 802.11 wireless networks operate. Earlier 802.11 interface controllers are usually only compatible with earlier variants of the standard, while newer cards support both current and old standards.

Specifications commonly used in marketing materials for WNICs include:

- Wireless data transfer rates (measured in Mbit/s); these range from 2 Mbit/s to 54 Mbit/s.

- Wireless transmit power (measured in dBm)

- Wireless network standards (may include standards such as 802.11b, 802.11g, 802.11n, etc.) 802.11g offers data transfer speeds equivalent to 802.11a – up to 54 Mbit/s – and the wider 300-foot (91 m) range of 802.11b, and is backward compatible with 802.11b.

Most Bluetooth cards do not implement any form of the 802.11 standard.

Range

Wireless range may be substantially affected by objects in the way of the signal and by the quality of the antenna. Large electrical appliances, such as refrigerators, fuse boxes, metal plumbing, and air conditioning units can impede a wireless network signal. The theoretical maximum range of IEEE 802.11 is only reached under ideal circumstances and true effective range is typically about half of the theoretical range. Specifically, the maximum throughput speed is only achieved at extremely close range (less than 25 feet (7.6 m) or so); at the outer reaches of a device's effective range, speed may decrease to around 1 Mbit/s before it drops out altogether. The reason is that wireless devices dynamically negotiate the top speed at which they can communicate without dropping too many data packets.

FullMAC and SoftMAC Devices

In an 802.11 WNIC, the *MAC Sublayer Management Entity* (MLME) can be implemented either in the NIC's hardware or firmware, or in host-based software that is executed on the main CPU. A WNIC that implements the MLME function in hardware or firmware is called a *FullMAC* WNIC or a *HardMAC* NIC and a NIC that implements it in host software is called a *SoftMAC* NIC.

A FullMAC device hides the complexity of the 802.11 protocol from the main CPU, instead providing an 802.3 (Ethernet) interface; a SoftMAC design implements only the timing-critical part of the protocol in hardware/firmware and the rest on the host.

FullMAC chips are typically used in mobile devices because:

- they are easier to integrate in complete products

- power is saved by having a specialized CPU perform the 802.11 processing;

- the chip vendor has tighter control of the MLME.

Popular example of FullMAC chips is the one implemented on the Raspberry Pi 3.

Linux kernel's *mac80211* framework provides capabilities for SoftMAC devices and additional capabilities (such as mesh networking, which is known as the IEEE 802.11s standard) for devices with limited functionality.

FreeBSD also supports SoftMAC drivers.

Wireless Distribution System

WDS stands for Wireless Distribution System and is a feature supported by an increasing number of 802.11 access points. Simply put, it enables single-radio APs to be wirelessly connected instead of using a wired Ethernet connection.

WDS connections are MAC address-based and employ a special Data Frame type that uses all four of the (MAC) address fields allowed in the 802.11 standard, instead of the three addresses used in normal AP <-> STA traffic.

The provision for four MAC addresses in a frame is about the only thing covered by the 802.11 standards, but it was enough to allow bridging features to first be added to enterprise-grade, i.e. expensive, 802.11b products in the late 1990's. Many of these implementations were based around a medium access control (MAC) layer design originated by a company called Choice Microsystems.

APs with wireless bridging features remained as high-priced items until fall 2002 when wireless bridging moved into consumer priced wireless products. D-Link first broke the artificially high wireless bridging price barrier by releasing a free upgrade to its DWL-900AP+ Access Point. This upgrade created the first consumer-priced

WLAN product to support bridging and repeating . Other companies soon followed with similar upgrades, and also introduced dedicated Wireless Bridges, such as Linksys' WET11 .

Though these products were actually making use of the WDS feature, they didn't refer to it as such. It wasn't until products based on Broadcom's 802.11g chipset started to hit the market at the beginning of 2003 that the WDS term started to be commonly used. Broadcom apparently included WDS support in its AP reference design code and WDS-enabled 802.11g APs have since become more widely available.

Technical

WDS may provide two modes of access point-to-access point (AP-to-AP) connectivity:

- Wireless bridging, in which WDS APs (AP-to-AP on sitecom routers AP) communicate only with each other and don't allow wireless stations (STA) (also known as wireless clients) to access them

- Wireless repeating, in which APs (WDS on sitecom routers) communicate with each other and with wireless STAs

In WDS mode, the two bits in *Frame Control header* of a wireless data frame called To DS and From DS are both set to 1. In this case, all four MAC addresses (Receiver's Address [RA], Transmitter's Address [TA], Source Address [SA] and Destination Address [DA]) are needed to successfully deliver the frame to its final destination. The TA is the address of the transmitting AP, while the RA is the address of receiving AP and stay the same as long as the configuration does not change.

Two disadvantages to using WDS are:

- The maximum wireless effective throughput may be halved after the first retransmission (hop) being made. For example, in the case of two APs connected via WDS, and communication is made between a computer which is plugged into the Ethernet port of AP A and a laptop which is connected wirelessly to AP B. The throughput is halved, because AP B has to retransmit the information during the communication of the two sides. However, in the case of communications between a computer which is plugged into the Ethernet port of AP A and a computer which is plugged into the Ethernet port of AP B, the throughput is not halved since there is no need to retransmit the information. Dual band/ radio APs may avoid this problem, by connecting to clients on one band/radio, and making a WDS network link with the other.

- Dynamically assigned and rotated encryption keys are usually not supported in a WDS connection. This means that dynamic Wi-Fi Protected Access (WPA) and other dynamic key assignment technology in most cases cannot be used, though WPA using pre-shared keys is possible. This is due to the lack of standardization in this field, which may be resolved with the upcoming 802.11s standard. As a result, only static WEP or WPA keys may be used in a WDS connection, including any STAs that associate to a WDS repeating AP.

OpenWrt, a universal third party router firmware, supports WDS with WPA-PSK, WPA2-PSK, WPA-PSK/WPA2-PSK Mixed-Mode encryption modes. Recent Apple base stations allow WDS with WPA, though in some cases firmware updates are required. Firmware for the Renasis SAP36g super access point and most third party firmware for the Linksys WRT54G(S)/GL support AES encryption using WPA2-PSK mixed-mode security, and TKIP encryption using WPA-PSK, while operating in WDS mode. However, this mode may not be compatible with other units running stock or alternate firmware.

Example

Suppose you have a WiFi-capable game console. This device needs to send one packet to a WAN host, and get one packet in reply.

Network 1: A wireless base station acting as a simple (non-WDS) wireless router. The packet leaves the game console, goes over-the-air to the router, which then transmits it across the WAN. One packet comes back, through the router, which transmits it wirelessly to the game console. Total packets sent over-the-air: 2.

Network 2: Two wireless base stations employing WDS: WAN connects to the master base station, that connects over-the-air to the remote base station, which talks over-the-air to the game console. The game console sends one packet over-the-air to the remote, which forwards it over-the-air to the master, which sends it to the WAN. Reply comes from the WAN to the master base station, over-the-air to the remote, and then over-the-air again to the game console. Total packets sent over-the-air: 4.

Network 3: Two wireless base stations employing WDS, but this time the game console connects by Ethernet cable to the remote base station. One packet goes from the game console over cable to the remote, from there by air to the master, and on to the WAN. Reply comes from WAN to master, over-the-air to remote, over cable to game console. Total packets sent over-the-air: 2.

Notice that network 1 (non-WDS) and network 3 (WDS) send the same number of packets over-the-air. The only slowdown is the potential halving due to the half-duplex nature of wifi.

But network 2 gets an additional halving because the remote base station uses double the air time because it's retransmitting over-the-air packets that it has just received over-the-air. This is the halving that is usually attributed to WDS, but that halving only happens when the route through a base station uses over-the-air links on both sides of it. That does not always happen in a WDS, and can happen in non-WDS.

Important Note: This "double hop" (one wireless hop from the main station to the remote station, and a second hop from the remote station to the wireless client [game console]) is not necessarily twice as slow. End to end latency introduced here is in the "store and forward" delay associated with the remote station forwarding packets. In order to accurately identify the true latency contribution of relaying through a wireless remote station vs. simply increasing the broadcast power of the main station, more comprehensive tests specific to the environment would be required.

Cracking of Wireless Networks

The security of wifi connections has been in and out of the news over the past few years as the integrity of the wifi encryption process has been progressively eroded. Wifi encryption is normally driven by the use of three flavours of passwords/passphrases – Wired Equivalent Privacy (WEP), Wifi Protected Access (WPA) and WPA2 – which use different methodologies to ensure (to differing degrees) the integrity of the wifi IP-based communications path. But all have come under attack, with tools available to intercept and crack authentication.

Cracking is the process of exploiting security weaknesses in wireless networks and gaining unauthorized access.

It is possible to crack the WEP/WPA keys used to gain access to a wireless network. Doing so requires software and hardware resources, and patience. The success of such attacks can also depend on how active and inactive the users of the target network are.

Wireless Network Basics

- Wireless local-area networks are based on IEEE 802.11. This is a set of standards defined by the Institute of Electrical and Electronics Engineers.

- 802.11 networks are either *infrastructure* networks or *ad hoc* networks. By default, people refer to infrastructure networks. Infrastructure networks are composed of one or more *access points* that coordinate the wireless traffic between the nodes and often connect the nodes to a wired network, acting as a bridge or a router.

  o Each access point constitutes a network that is named a *basic service set* or BSS. A BSS is identified by a BSSID, usually the MAC address of the access point.

  o Each access point is part of an *extended service set* or ESS, which is identified by an ESSID or SSID in short, usually a character string.

  o A basic service set consists of one access point and several wireless *clients*.

An extended service set is a configuration with multiple access points and roaming capabilities for the clients. An independent basic service set or IBSS is the ad hoc configuration. This configuration allows wireless clients to connect to each other directly, without an access point as a central manager.

o Access points broadcast a signal regularly to make the network known to clients. They relay traffic from one wireless client to another. Access points may determine which clients may connect, and when clients do, they are said to be *associated* with the access point. To obtain access to an access point, both the BSSID and the SSID are required.

- Ad hoc networks have no access point for central coordination. Each node connects in a peer-to-peer way. This configuration is an *independent basic service set* or IBSS. Ad hoc networks also have an SSID.

Wireless Network Frames

802.11 networks use *data frames, management frames,* and *control frames.* Data frames convey the real data, and are similar to those of Ethernet. Management frames maintain both network configuration and connectivity. Control frames manage access to the ether and prevent access points and clients from interfering with each other in the ether. Some information on management frames will be helpful to better understand what programs for *reconnaissance* do.

- *Beacon frames* are used primarily in reconnaissance. They advertise the existence and basic configuration of the network. Each frame contains the BSSID, the SSID, and some information on basic authentication and encryption. Clients use the flow of beacon frames to monitor the signal strength of their access point.

- *Probe request frames* are almost the same as the beacon frames. A probe request frame is sent from a client when it wants to connect to a wireless network. It contains information about the requested network.

- *Probe response frames* are sent to clients to answer probe request frames. One response frame answers each request frame, and it contains information on the capabilities and configurations of the network. Useful for reconnaissance.

- *Authentication request frames* are sent by clients when they want to connect to a network. Authentication precedes association in infrastructure networks. Either *open* authentication or *shared key* authentication is possible. After serious flaws were found in shared key authentication, most networks switched to open authentication, combined with a stronger authentication method applied after the association phase.

- *Authentication response frames* are sent to clients to answer authentication request frames. There is one answer to each request, and it contains either status information or a challenge related to shared key authentication.

- *Association request frames* are sent by clients to associate with the network. An association request frame contains much of the same information as the probe request contains, and it must have the SSID. This can be used to obtain the SSID when a network is configured to hide the SSID in beacon frames.

- *Association response frames* are sent to clients to answer an association request frame. They contain a bit of network information and indicate whether the association was successful.

- *Deauthentication* and *disassociation frames* are sent to a node to notify that an authentication or an association has failed and must be established anew.

Reconnaissance of Wireless Networks

Wardriving is a common method of wireless network reconnaissance. A well-equipped wardriver uses a laptop computer with a wireless card, an antenna mounted on the car, a power inverter, a connected GPS receiver, and a way to connect to the internet wirelessly. The purpose of wardriving is to locate a wireless network and to collect information about its configuration and associated clients.

The laptop computer and the wireless card must support a mode called *monitor* or *rfmon*.

Netstumbler

Netstumbler is a network discovery program for Windows. It is free. Netstumbler has become one of the most popular programs for wardriving and wireless reconnaissance, although it has a disadvantage. It can be detected easily by most wireless intrusion detection systems, because it actively probes a network to collect information. Netstumbler has integrated support for a GPS unit. With this support, Netstumbler displays GPS coordinate information next to the information about each discovered network, which can be useful for finding specific networks again after having sorted out collected data.

The latest release of Netstumbler is of 1 April 2004. It does not work well with 64-bit Windows XP or Windows Vista.

inSSIDer

inSSIDer is a Wi-Fi network scanner for the 32-bit and 64-bit versions of Windows XP, Vista, 7, Windows 8 and Android. It is free and open source. The software uses the current wireless card or a wireless USB adapter and supports most GPS devices (namely those that use NMEA 2.3 or higher). Its graphical user interface shows MAC address, SSID, signal strength, hardware brand, security, and network type of nearby Wi-Fi networks. It can also track the strength of the signals and show them in a time graph.

Kismet

Kismet is a wireless network traffic analyser for OS X, Linux, OpenBSD, NetBSD, and FreeBSD. It is free and open source. Kismet has become the most popular program for serious wardrivers. It offers a rich set of features, including deep analysis of captured traffic.

Wireshark

Wireshark is a packet sniffer and network traffic analyser that can run on all popular operating systems, but support for the capture of wireless traffic is limited. It is free and open source. Decoding and analysing wireless traffic is not the foremost function of Wireshark, but it can give results that cannot be obtained with other programs. Wireshark requires sufficient knowledge of the network protocols to obtain a full analysis of the traffic, however.

Analysers of AirMagnet

AirMagnet Laptop Analyser and AirMagnet Handheld Analyser are wireless network analysis tools made by AirMagnet. The company started with the Handheld Analyser, which was very suitable for surveying sites where wireless networks were deployed as well as for finding rogue access points. The Laptop Analyser was released because the hand-held product was impractical for the reconnaissance of wide areas. These commercial analysers probably offer the best combination of powerful analysis and simple user interface. However, they are not as well adapted to the needs of a wardriver as some of the free programs.

Androdumpper

Androdumpper is an Android APK that is used to test and hack WPS Wireless routers which have a vulnerability by using algorithms to hack into that WIFI network. It runs best on Android version 5.0+

Airopeek

Airopeek is a packet sniffer and network traffic analyser made by Wildpackets. This commercial program supports Windows and works with most wireless network interface cards. It has become the industrial standard for capturing and analysing wireless traffic. However, like Wireshark, Airopeek requires thorough knowledge of the protocols to use it to its ability.

KisMac

KisMac is a program for the discovery of wireless networks that runs on the OS X operating system. The functionality of KisMac includes GPS support with mapping, SSID decloaking, deauthentication attacks, and WEP cracking.

Penetration of a Wireless Network

There are two basic types of vulnerabilities associated with WLANs: those caused by poor configuration and those caused by poor encryption. Poor configuration causes many vulnerabilities. Wireless networks are often put into use with no or insufficient security settings. With no security settings – the default configuration – access is obtained simply by association. Without sufficient security settings, networks can easily be defeated by cloaking and/or MAC address filtering. Poor encryption causes the remaining vulnerabilities. Wired Equivalent Privacy (WEP) is defective and can be defeated in several ways. Wi-Fi Protected Access (WPA) and Cisco's Lightweight Extensible Authentication Protocol (LEAP) are vulnerable to dictionary attacks.

Encryption Types and their Attacks

Wired Equivalent Privacy (WEP)

WEP was the encryption standard firstly available for wireless networks. It can be deployed in 64 and 128 bit strength. 64 bit WEP has a secret key of 40 bits and an initialisation vector of 24 bits, and is often called 40 bit WEP. 128 bit WEP has a secret key of 104 bits and an initialisation vector of 24 bits, and is called 104 bit WEP. Association is possible using a password, an ASCII key, or a hexadecimal key. There are two methods for cracking WEP: the *FMS attack* and the *chopping attack*. The FMS attack – named after Fluhrer, Mantin, and Shamir – is based on a weakness of the RC4 encryption algorithm . The researchers found that 9000 of the possible 16 million initialisation vectors can be considered weak, and collecting enough of them allows the determination of the encryption key. To crack the WEP key in most cases, 5 million encrypted packets must be captured to collect about 3000 weak initialisation vectors. (In some cases 1500 vectors will do, in some other cases more than 5000 are needed for success.) The weak initialisation vectors are supplied to the Key Scheduling Algorithm (KSA) and the Pseudo Random Generator (PRNG) to determine the first byte of the WEP key. This procedure is then repeated for the remaining bytes of the key. The chopping attack chops the last byte off from the captured encrypted packets. This breaks the Cyclic Redundancy Check/Integrity Check Value (CRC/ICV). When all 8 bits of the removed byte were zero, the CRC of the shortened packet is made valid again by manipulation of the last four bytes. This manipulation is: result = original XOR certain value. The manipulated packet can then be retransmitted. This method enables the determination of the key by collecting *unique* initialisation vectors. The main problem with both the FMS attack and the chopping attack is that capturing enough packets can take weeks or sometimes months. Fortunately, the speed of capturing packets can be increased by injecting packets into the network. One or more Address Resolution Protocol (ARP) packets are usually collected to this end, and then transmitted to the access point repeatedly until enough response packets have been captured. ARP packets are a good choice because they have a recognizable size of 28 bytes. Waiting for a legitimate ARP packet can take awhile. ARP packets are most commonly transmitted during an authentication process. Rather than waiting for that, sending a deauthentication frame that pushes a client off the network will require that client to reauthenticate. This often creates an ARP packet.

Wi-Fi Protected Access (WPA/WPA2)

WPA was developed because of the vulnerabilities of WEP. WPA uses either a pre-shared key (WPA-PSK) or is used in combination with a RADIUS server (WPA-RADIUS). For its encryption algorithm, WPA uses either the Temporal Key Integrity Protocol (TKIP) or the Advanced Encryption Standard (AES). WPA2 was developed because of some vulnerabilities of WPA-PSK and to strengthen the encryption further. WPA2 uses both TKIP and AES, and requires not only an encryption piece but also an authentication piece. A form of the Extensible Authentication Protocol (EAP) is deployed for this piece. WPA-PSK can be attacked when the PSK is shorter than 21 characters. Firstly, the four-way EAP Over LAN (EAPOL) handshake must be captured. This can be captured during a legitimate authentication, or a reauthentication can be forced by sending deauthentication packets to clients. Secondly, each word of a word-list must be hashed with the Hashed Message Authentication Code – Secure Hash Algorithm 1 and two so called nonce values, along with the MAC address of the client that asked for authentication and the MAC address of the access point that gave authentication. Word-lists can be found at. LEAP uses a variation of Microsoft Challenge Handshake Protocol version 2 (MS-CHAPv2). This handshake uses the Data Encryption Standard (DES) for key selection. LEAP can be cracked with a dictionary attack. The attack involves capturing an authentication sequence and then comparing the last two bytes of a captured response with those generated with a word-list. WPA-RADIUS cannot be cracked. However, if the RADIUS authentication server itself can be cracked, then the whole network is imperilled. The security of authentication servers is often neglected. WPA2 can be attacked by using the WPA-PSK attack, but is largely ineffective.

Aircrack-ng

Aircrack-ng runs on Windows and Linux, and can crack WEP and WPA-PSK. It can use the Pychkine-Tews-Weinmann and KoreK attacks, both are statistical methods that are more efficient than the traditional FMS attack. Aircrack-ng consists of components. Airmon-ng configures the wireless network card. Airodump-ng captures the frames. Aireplay-ng generates traffic. Aircrack-ng does the cracking, using the data collected by airodump-ng. Finally, airdecap-ng decrypts all packets that were captured. Thus, aircrack-ng is the name of the suite and also of one of the components.

CoWPAtty

CoWPAtty automates the dictionary attack for WPA-PSK. It runs on Linux. The program is started using a command-line interface, specifying a word-list that contains the passphrase, a dump file that contains the four-way EAPOL handshake, and the SSID of the network.

Void11

Void11 is a program that deauthenticates clients. It runs on Linux.

MAC Address Filtering and its Attack

MAC address filtering can be used alone as an ineffective security measure, or in combination with encryption. The attack is determining an allowed MAC address, and then changing the MAC address of the attacker to that address. EtherChange is one of the many programs available to change the MAC address of network adapters. It runs on Windows.

In conclusion Penetration testing of a wireless network is often a stepping stone for penetration testing of the internal network. The wireless network then serves as a so-called *entry vector*. If WPA-RADIUS is in use at a target site, another entry vector must be investigated.

Reconnaissance of the Local-area Network

Sniffing

A 'wireless' sniffer can find IP addresses, which is *helpful* for network mapping.

Access points usually connect the nodes of a wireless network to a wired network as a bridge or a router. Both a bridge and a router use a routing table to forward packets.

Footprinting

Finding *relevant* and *reachable* IP addresses is the objective of the reconnaissance phase of attacking an organization over the Internet. The relevant IP addresses are determined by collecting as many DNS host names as possible and translating them to IP addresses and IP address ranges. This is called footprinting.

A search engine is the key for finding as much information as possible about a target. In many cases, organizations do not want to protect all their resources from internet access. For instance, a web server must be accessible. Many organizations additionally have email servers, FTP servers, and other systems that must be accessible over the internet. The IP addresses of an organization are often grouped together. *If one IP address has been found, the rest probably can be found around it.*

Name servers store tables that show how domain names must be translated to IP addresses and vice versa. With Windows, the command NSLookup can be used to query DNS servers. When the word help is entered at NSLookup's prompt, a list of all commands is given. With Linux, the command dig can be used to query DNS servers. It displays a list of options when invoked with the option -h only. And the command host reverses IP addresses to hostnames. The program nmap can be used as a reverse DNS walker: nmap -sL 1.1.1.1-30 gives the reverse entries for the given range.

ARIN, RIPE, APNIC, LACNIC, and AFRINIC are the five Regional Internet Registries that are responsible for the assignment and registration of IP addresses. All have a website with which their databases can be searched for the owner of an IP address. Some of the Registries respond to a search for the name of an organization with a list of all IP address ranges that are assigned to the name. However, the records of the Registries are not always correct and are in most cases useless.

Probably most computers with access to the internet receive their IP address dynamically by DHCP. This protocol has become more popular over the last years because of a decrease of available IP addresses and an increase of large networks that are dynamic. DHCP is particularly important when many employees take a portable computer from one office to another. The router/firewall device that people use at home to connect to the internet probably also functions as a DHCP server.

Nowadays many router/DHCP devices perform Network Address Translation (NAT). The NAT device is a gateway between the local network and the internet. Seen from the internet, the NAT device seems to be a single *host*. With NAT, the local network can use any IP address space. Some IP address ranges are reserved for private networks. These ranges are typically used for the local area network behind a NAT device, and they are: 10.0.0.0 - 10.255.255.255, 172.16.0.0 - 172.31.255.255, and 192.168.0.0 - 192.168.255.255.

The relevant IP addresses must be narrowed down to those that are reachable. For this purpose, the process of scanning enters on the scene.

Host Scanning

Once access to a wireless network has been gained, it is helpful to determine the network's topology, including the names of the computers connected to the network. Nmap can be used for this, which is available in a Windows and a Linux version. However, Nmap does not provide the user with a network diagram. The network scanner Network View that runs on Windows does. *The program asks for one IP address or an IP address range.* When the program has finished scanning, it displays a map of the network using different pictures for routers, workstations, servers, and laptops, all with their names added.

The most direct method for finding hosts on a LAN is using the program ping. When using a modern flavour of Unix, shell commands can be combined to produce custom ping-sweeps. When using Windows, the command-line can also be used to create a ping-sweep.

Ping-sweeps are also known as *host scans*. Nmap can be used for a host scan when the option -sP is added: nmap -n -sP 10.160.9.1-30 scans the first 30 addresses of the subnet 10.160.9, where the -n option prevents reverse DNS lookups.

Ping packets could reliably determine whether a computer was on line at a specified IP address. Nowadays these ICMP echo request packets are sometimes blocked by the firewall of an operating system. Although Nmap also probes TCP port 80, specifying more TCP ports to probe is recommended when pings are blocked. Consequently, nmap -sP -PS21,22,23,25,80,139,445,3389 10.160.9.1-30 can achieve better results. And by combining various options as in nmap -sP -PS21,22,23,25,80,135,139,445,1025,3389 -PU53,67,68,69,111,161,445,514 -PE -PP -PM 10.160.9.1-30, superb host scanning is achieved.

Nmap is available for Windows and most Unix operating systems, and offers graphical and command-line interfaces.

Port Scanning

The purpose of port scanning is finding the *open ports* on the computers that were found with a host scan. When a port scan is started on a network without making use of the results of a host scan, much time is wasted when many IP addresses in the address range are vacant.

Open Ports

Most programs that communicate over the internet use either the TCP or the UDP protocol. Both protocols support 65536 so called *ports* that programs can choose to bind to. This allows programs to run concurrently on one IP address. Most programs have default ports that are most often used. For example, HTTP servers commonly use TCP port 80.

Network scanners try to connect to TCP or UDP ports. When a port accepts a connection, it can be assumed that the commonly bound program is running.

TCP connections begin with a SYN packet being sent from client to server. The server responds with a SYN/ACK packet. Finally, the client sends an ACK packet. When the scanner sends a SYN packet and gets the SYN/ACK packet back, the port is considered open. When a RST packet is received instead, the port is considered closed. When no response is received the port is either considered filtered by a firewall or there is no running host at the IP address.

Scanning UDP ports is more difficult because UDP does not use handshakes and programs tend to discard UDP packets that they cannot process. When an UDP packet is sent to a port that has no program bound to it, an ICMP error packet is returned. That port can then be considered closed. When no answer is received, the port can be considered either filtered by a firewall or open. Many people abandoned UDP scanning because simple UDP scanners cannot distinguish between filtered and open ports.

Common Ports

Although it is most thorough to scan all 65536 ports, this would take more time than scanning only the most common ports. Therefore, Nmap scans 1667 TCP ports by default (in 2007).

Specifying Ports

The -p option instructs Nmap to scan specified ports, as in nmap -p 21-25,80,100-160 10.150.9.46. Specifying TCP and UDP ports is also possible, as in nmap -pT:21-25,80,U:5000-5500 10.150.9.46.

Specifying Targets

Nmap always requires the specification of a host or hosts to scan. A single host can be specified with an IP address or a domain name. Multiple hosts can be specified with IP address ranges. Examples are 1.1.1.1, www.company.com, and 10.1.50.1-5,250-254.

Specifying Scan Type

TCP SYN scan

Nmap performs a TCP SYN scan by default. In this scan, the packets have only their SYN flag set. The -sS option specifies the default explicitly. When Nmap is started with administrator privileges, this default scan takes effect. When Nmap is started with user privileges, a connect scan is performed.

TCP Connect Scan

The -sT option instructs Nmap to establish a full connection. This scan is inferior to the previous because an additional packet must be sent and logging by the target is more likely. The connect scan is performed when Nmap is executed with user privileges or when IPv6 addresses are scanned.

TCP Null Scan

The -sN option instructs Nmap to send packets that have none of the SYN, RST, and ACK flags set. When the TCP port is closed, a RST packet is sent in return. When the

TCP port is open or filtered, there is no response. The null scan can often bypass a stateless firewall, but is not useful when a stateful firewall is employed.

UDP Empty Packet Scan

The -sU option instructs Nmap to send UDP packets with no data. When an ICMP error is returned, the port can be assumed closed. When no response is received, the port can be assumed open or filtered. No differentiation between open and filtered ports is a severe limitation.

UDP Application Data Scan

The -sU -sV options instruct Nmap to use application data for application identification. This combination of options can lead to very slow scanning.

Specifying Scan Speed

When packets are sent to a network faster than it can cope with they will be dropped. This leads to inaccurate scanning results. When an intrusion detection system or intrusion prevention system is present on the target network, detection becomes more likely as speed increases. Many IPS devices and firewalls respond to a storm of SYN packets by enabling SYN cookies that make appear every port to be open. Full speed scans can even wreak havoc on stateful network devices.

Nmap provides five templates for adjusting speed and also adapts itself. The -T0 option makes it wait for 5 minutes before the next packet is sent, the -T1 option makes it wait for 15 seconds, -T2 inserts 0.4 seconds, -T3 is the default (which leaves timing settings unchanged), -T4 reduces time-outs and retransmissions to speed things up slightly, and -T5 reduces time-outs and retransmissions even more to speed things up significantly. Modern IDS/IPS devices can detect scans that use the -T1 option. The user can also define a new template of settings and use it instead of a provided one.

Application Identification

The -sV option instructs Nmap to also determine the version of a running application.

Operating System Identification

The -O option instructs Nmap to try to determine the operating systems of the targets. Specially crafted packets are sent to open and closed ports and the responses are compared with a database.

Saving Output

The -oX <filename> option instructs Nmap to save the output to a file in XML format.

Vulnerability Scanning

A vulnerability is a bug in an application program that affects security. They are made public on places such as the BugTraq and the Full-Disclosure mailing lists. The Computer Emergency Response Team (CERT) brings out a statistical report every year. There were 8064 vulnerabilities counted in 2006 alone.

Vulnerability scanning is determining whether known vulnerabilities are present on a target.

Nessus

Nessus is probably the best known vulnerability scanner. It is free and has versions for Windows, OS X, Linux, and FreeBSD. Nessus uses plug-ins to find vulnerabilities by sort. Updated plug-ins are regularly released.

Nessus offers a non-intrusive scan, an intrusive scan that can harm the target, and a custom scan. A scan requires the IP addresses or domain names of the targets. Nessus begins with a port scan to identify the programs that are running and the operating systems of the targets. It ends with a report that specifies all open ports and their associated vulnerabilities.

Nikto

Nikto is a web scanner that can identify vulnerable applications and dangerous files. It is open source software and has versions for Windows and Linux. The program uses a command-line interface of the operating system.

Exploitation of a Vulnerability

An exploit takes advantage of a bug in an application. This can take effect in the execution of arbitrary commands by inserting them in the execution path of the program. Escalation of privileges, bypass of authentication, or infringement of confidentiality can be the result.

Metasploit

The Metasploit framework was released in 2003. This framework provided for the first time:

- a single *exploit* database with easy updating,
- freely combining of an exploit with a *payload*,
- a consistent interface for setting *options*, and
- integrated *encoding* and *evasion*,

where:

- an exploit is a code module that uses a particular vulnerability,

- a payload is code that is sent along with the exploit to take some action, such as providing a command-line interface,

- options are used to select variants of exploits and payloads,

- encoding is modifying the payload to circumvent limitations, whether they are caused by the logic of the vulnerability or an inadequate IPS, and

- evasion is bypassing security devices by employing evasion techniques.

The basic procedure of using Metasploit is: choose an exploit, choose a payload, set the IP address and port of the target, start the exploit, evaluate, and stop or repeat the procedure.

Metasploit is not suited for finding the vulnerabilities of a host; a vulnerability scanner is. Alternatively, when a port scanner has found an open port, all exploits for that port may be tried.

Metasploit 3.0 provides the following payloads:

- VNC injection. This payload for targets that run Windows gives a graphical user interface to the target that is synchronized with the graphical user interface of the target.

- File execution. This payload for targets that run Windows uploads a file and executes it.

- Interactive shell. This payload gives a command-line interface to the target.

- Add user. This payload adds a user with specified name and password that has administrator access.

- Meterpreter. This payload gives a rich command-line interface to targets that run Windows.

VNC connections need a relatively large bandwidth to be usable, and if someone is in front of the compromised computer then any interaction will be seen very quickly. The command-line interfaces of Linux and OS X are powerful, but that of Windows is not. The Meterpreter payload remedies these shortcomings. The reference gives a list of Meterpreter commands.

Maintaining Control

The ultimate gratification for a network intruder always is to obtain administrator privileges for a network. When an intruder is inside, one of his or her first undertakings is often to install a so-called *rootkit* on a target computer. This is a collection of programs to facilitate durable influence on a system. Some of these programs are used to compromise new user accounts or new computers on the network. Other programs are to obscure the presence of the intruder. These obscuring programs may include false versions of standard network utilities such as netstat, or programs that can remove all data from the log files of a computer that relate to the intruder. Yet other programs of a rootkit may be used to survey the network or to overhear more passwords that are travelling over it. Rootkits may also give the means to change the very operating system of the computer it is installed on.

The network intruder then proceeds with creating one or more so called *back doors*. These are access provisions that are hard to find for system administrators, and they serve to prevent the logging and monitoring that results from normal use of the network. A back door may be a concealed account or an account of which the privileges have been escalated. Or it may be a utility for remote access, such as Telnet, that has been configured to operate with a port number that is not customary.

The network intruder then proceeds with stealing files, or stealing credit card information, or preparing a computer to send spam emails at will. Another goal is to prepare for the next intrusion. A cautious intruder is protective against discovery of his or her location. The method of choice is to use a computer that already has been attacked as an intermediary. Some intruders use a series of intermediate computers, making it impracticable to locate them.

Back Doors

The purpose of a back door is to maintain a communication channel and having methods to control a host that has been gained entry to. These methods include those for file transfer and the execution of programs. It is often important to make sure that the access or communication remains secret. And access control is desirable in order to prevent others from using the back door.

Back Orifice 2000 was designed as a back door. The server runs on Windows, and there are clients for Windows, Linux and other operating systems. The server is configured easily with a utility. After configuration, the server needs to be uploaded to the target and then started. Back Orifice 2000 supports file transfer, file execution, logging of keystrokes, and control of connections. There is also an AES plug-in for traffic encryption and an STCPIO plug-in for further obfuscation of the traffic. The first plug-in adds security and the combination of these plug-ins makes it much harder for an IDS to relate the traffic to a back door.

Rootkits

Rootkits specialize in hiding themselves and other programs.

Hacker Defender (hxdef) is an open source rootkit for Windows. It can hide its files, its process, its registry entries, and its port in multiple DLLs. Although it has a simple command-line interface as a back door, it is often better to use its ability to hide a more appropriate tool.

Prevention and Protection

An unprotected wireless network is extremely insecure. From anywhere within broadcast range, someone can eavesdrop or start using the network. Therefore, the IEEE 802.11 standard for wireless networks was accompanied with Wired Equivalent Privacy (WEP). This security protocol takes care of the following:

- *authentication*: assurance that all participants are who they state they are, and are authorized to use the network

- *confidentiality*: protection against eavesdropping

- *integrity*: assurance of data being unaltered

WEP has been criticized by security experts. Most experts regard it as ineffective by now.

In 2004 a draft for a better security protocol appeared, and it was included in the IEEE 802.11 standard in 2007. This new protocol, WPA2, uses an AES block cipher instead of the RC4 algorithm and has better procedures for authentication and key distribution. WPA2 is much more secure than WEP, but WEP was still in wide use in 2009.

Many wireless routers also support controlling the MAC addresses of computers that are authorized to use a wireless network. This measure can effectively stop a neighbour from using the network, but experienced intruders will not be stopped. MAC filtering can be attacked because a MAC address can be faked easily.

In the past, turning off the broadcasting of the SSID has also been thought to give security to a wireless network. This is not the case however. Freely available tools exist that quickly discover an SSID that is not broadcast. Microsoft has also determined that switching off the broadcasting of the SSID leads to less security. Details can be found in *Non-broadcast Wireless Networks with Microsoft Windows*.

Returning to encryption, the WEP specification at any encryption strength is unable to withstand determined hacking. Therefore, Wi-Fi Protected Access (WPA) was derived from WEP. Software upgrades are often available. The latest devices that conform to the 802.11g or 802.11n standards also support WPA2. (WPA uses the TKIP encryption, WPA2 uses the stronger AES method.) It is recommended to use only hardware that supports WPA or WPA2.

Installing updates regularly, disabling WPS, setting a custom SSID, requiring WPA2, and using a strong password make a wireless router more difficult to crack. Even so, unpatched security flaws in a router's software or firmware may still be used by an attacker to bypass encryption and gain control of the device. Many router manufacturers do not always provide security updates in a timely manner, or at all, especially for more inexpensive models.

WPS currently has a severe vulnerability in which the 8 pin numbered (0-9) passwords being used can easily be split into two halves, this means that each half can be brute-forced individually and so the possible combinations are greatly lessened ($10^4 + 10^4$, as opposed to 10^8). This vulnerability has been addressed by most manufacturers these days by using a lock down mechanism where the router will automatically lock its WPS after a number of failed pin attempts (it can take a number of hours before the router will automatically unlock, some even have to be rebooted which can make WPS attacks completely obsolete). Without a lock down feature, a WPA2 router with WPS enabled can easily be cracked in 5 hours using a brute force WPS attack.

SSID's are used in routers not only to identify them within the mass of 2.4, 3.6, 5 and 60 GHz frequencies which are currently flying around our cities, but are also used as a "seed" for the router's password hashes. Standard and popular SSID's such as "Netgear" can be brute forced through the use of rainbow tables, however, it should be noted that the use of a salt greatly improves security against rainbow tables. The most popular method of WPA and WPA2 cracking is through obtaining what's known as a "4 way handshake". when a device is connecting with a network there is a 4-stage authorization process referred to as a 4 way handshake. When a wireless device undergoes this process this handshake is sent through the air and can easily be monitored and saved by an external system. The handshake will be encrypted by the router's password, this means that as opposed to communicating with the router directly (which can be quite slow), the cracker can attempt to brute force the handshake itself using dictionary attacks. A device that is connected directly with the router will still undergo this very process, however, the handshake will be sent through the connected wire as opposed to the air so it cannot be intercepted. If a 4 way handshake has already been intercepted, it does not mean that the cracker will be granted immediate access however. If the password used contains at least 12 characters consisting of both random upper and lower case letters and numbers that do not spell a word, name or have any pattern then the password will be essentially uncrackable. Just to give an example of this, let's just take the minimum of 8 characters for WPA2 and suppose we take upper case and lower case letters, digits from 0-9 and a small selection of symbols, we can avail of a hefty choice of 64 characters. In an 8 character length password this is a grand total of 64^8 possible combinations. Taking a single machine that could attempt 500 passwords per second, this gives us just about 17,900 years to attempt every possible combination. Not even to mention the amount of space necessary to store each combination in a dictionary.

Note: The use of MAC filtering to protect your network will not work as MACs using the network can be easily detected and spoofed.

Detection

A network scanner or sniffer is an application program that makes use of a wireless network interface card. It repeatedly tunes the wireless card successively to a number of radio channels. With a passive scanner this pertains only to the receiver of the wireless card, and therefore the scanning cannot be detected.

An attacker can obtain a considerable amount of information with a passive scanner, but more information may be obtained by sending crafted frames that provoke useful responses. This is called active scanning or probing. Active scanning also involves the use of the transmitter of the wireless card. The activity can therefore be detected and the wireless card can be located.

Detection is possible with an intrusion detection system for wireless networks, and locating is possible with suitable equipment.

Wireless intrusion detection systems are designed to detect anomalous behaviour. They have one or more sensors that collect SSIDs, radio channels, beacon intervals, encryption, MAC addresses, transmission speeds, and signal-to-noise ratios. Wireless intrusion detection systems maintain a registry of MAC addresses with which unknown clients are detected.

Crackers and Society

There is consensus that computer attackers can be divided in the following groups.

- Adolescent amateurs. They often have a basic knowledge of computer systems and apply scripts and techniques that are available on the internet.

- Adult amateurs. Most of them are motivated by the intellectual challenge.

- Professionals. They know much about computers. They are motivated by the financial reward but they are also fond of their activity.

Naming of Crackers

The term *hacker* was originally used for someone who could modify a computer for his or her own purposes. *Hacking* is an intrusion combined with direct alteration of the security or data structures of the breached system. The word hacking is often confused with cracking in popular media discourse, and obfuscates the fact that hacking is less about eavesdropping and more related to interference and alteration. However, because of the consistent abuse by the news media, in 2007 the term hacker was commonly used for someone who accesses a network or a computer without authorization of the owner.

In 2011, *Collins Dictionary* stated that the word hacker can mean a computer fanatic, in particular one who by means of a personal computer breaks into the computer system of a company, government, or the like. It also denoted that in that sense the word hacker is slang. Slang words are not appropriate in formal writing or speech.

Computer experts reserve the word hacker for a very clever programmer. They call someone who breaks into computers an intruder, attacker, or cracker.

Software

- Aircrack-ng

- BackTrack 5 – This latest release from Offensive Security is based on Ubuntu 10.04 LTS Linux. Three graphical desktop environments can be chosen from: Gnome, KDE, and Fluxbox. Over 300 application programs are included for penetration testing, such as network monitors and password crackers, but also Metasploit 3.7.0, an exploit framework. BackTrack 5 is a live distribution, but there is also an ARM version available for the Android operating system, allowing tablets and smartphones to be used for mobile penetration testing of Wi-Fi networks. BackTrack can be installed on hard disk, both alone and in dual boot configuration, on a USB flash drive, and in VMware. Metasploit's effectiveness is caused by the large number of exploits that are updated continually. In August 2011, there were 716 exploits for all usual operating systems together. Armitage is the GUI for Metasploit within BackTrack 5. This GUI can import files in XML format, and it supports Nmap, Nessus, and Metasploit-Express.

How to Secure Wireless Networks

In minimizing wireless network attacks; an organization can adopt the following policies

- Changing default passwords that come with the hardware

- Enabling the authentication mechanism

- Access to the network can be restricted by allowing only registered MAC addresses.

- Use of strong WEP and WPA-PSK keys, a combination of symbols, number and characters reduce the chance of the keys been cracking using dictionary and brute force attacks.

- Firewall Software can also help reduce unauthorized access.

References

- Guowang Miao, Jens Zander, Ki Won Sung, and Ben Slimane, Fundamentals of Mobile Data Networks, Cambridge University Press, ISBN 1107143217, 2016

- Foster, Kenneth R (March 2007). "Radiofrequency exposure from wireless LANs utilizing Wi-Fi technology". Health Physics. 92 (3): 280–289. doi:10.1097/01.HP.0000248117.74843.34. PMI D 17293700

- Team Digit (Jan 2006). "Fast Track to Mobile Telephony". Internet Archive. Jasubhai Digital Media. Archived from the original (text) on 8 June 2014. Retrieved 1 March 2017

- Ermanno Pietrosemoli. "Setting Long Distance WiFi Records: Proofing Solutions for Rural Connectivity". Fundación Escuela Latinoamericana de Redes University of the Andes (Venezuela). Retrieved March 17, 2012

- Sharma, Chetan; Nakamura, Yasuhisa (2003-11-20). Wireless Data Services: Technologies, Business Models and Global Markets. Cambridge University Press. ISBN 978-0-521-82843-7

2

Types of Wireless Networks

There are different categories of wireless networks, depending on the number of devices that a network can connect, or the extent of its functioning. These are Wireless LAN, WAN, PAN, MAN, cellular network, etc., which have been discussed in great detail in this chapter.

Wireless LAN

A wireless local area network (WLAN) is a wireless distribution method for two or more devices that use high-frequency radio waves and often include an access point to the Internet. A WLAN allows users to move around the coverage area, often a home or small office, while maintaining a network connection.

A WLAN is sometimes call a local area wireless network (LAWN).

In the early 1990s, WLANs were very expensive and were only used when wired connections were strategically impossible. By the late 1990s, most WLAN solutions and proprietary protocols were replaced by IEEE 802.11 standards in various versions (versions "a" through "n"). WLAN prices also began to decrease significantly.

WLAN should not be confused with the Wi-Fi Alliance's Wi-Fi trademark. Wi-Fi is not a technical term, but is described as a superset of the IEEE 802.11 standard and is sometimes used interchangeably with that standard. However, not every Wi-Fi device actually receives Wi-Fi Alliance certification, although Wi-Fi is used by more than 700 million people through about 750,000 Internet connection hot spots.

Every component that connects to a WLAN is considered a station and falls into one of two categories: access points (APs) and clients. APs transmit and receive radio frequency signals with devices able to receive transmitted signals; they normally function as routers. Clients may include a variety of devices such as desktop computers, workstations, laptop computers, IP phones and other cell phones and Smartphones. All stations able to communicate with each other are called basic service sets (BSSs), of which there are two types: independent and infrastructure. Independent BSSs (IBSS) exist when two clients communicate without using APs, but cannot connect to any other BSS. Such WLANs are called a peer-to-peer or an ad-hoc WLANs. The second BSS is called an infrastructure BSS. It may communicate with other stations but only in other BSSs and it must use APs.

Architecture

Stations

All components that can connect into a wireless medium in a network are referred to as stations (STA). All stations are equipped with wireless network interface controllers (WNICs). Wireless stations fall into two categories: wireless access points, and clients. Access points (APs), normally wireless routers, are base stations for the wireless network. They transmit and receive radio frequencies for wireless enabled devices to communicate with. Wireless clients can be mobile devices such as laptops, personal digital assistants, IP phones and other smartphones, or non-portable devices such as desktop computers, printers, and workstations that are equipped with a wireless network interface.

Basic Service Set

The basic service set (BSS) is a set of all stations that can communicate with each other at PHY layer. Every BSS has an identification (ID) called the BSSID, which is the MAC address of the access point servicing the BSS.

There are two types of BSS: Independent BSS (also referred to as IBSS), and infrastructure BSS. An independent BSS (IBSS) is an ad hoc network that contains no access points, which means they cannot connect to any other basic service set.

Independent Basic Service Set

An IBSS is a set of STAs configured in ad hoc (peer-to-peer)mode.

Extended Service Set

An extended service set (ESS) is a set of connected BSSs. Access points in an ESS are connected by a distribution system. Each ESS has an ID called the SSID which is a 32-byte (maximum) character string.

Distribution System

A distribution system (DS) connects access points in an extended service set. The concept of a DS can be used to increase network coverage through roaming between cells.

DS can be wired or wireless. Current wireless distribution systems are mostly based on WDS or MESH protocols, though other systems are in use.

Types of Wireless LANs

The IEEE 802.11 has two basic modes of operation: infrastructure and *ad hoc* mode. In *ad hoc* mode, mobile units transmit directly peer-to-peer. In infrastructure mode, mobile units communicate through an access point that serves as a bridge to other networks (such as Internet or LAN).

Since wireless communication uses a more open medium for communication in comparison to wired LANs, the 802.11 designers also included encryption mechanisms: Wired Equivalent Privacy (WEP, now insecure), Wi-Fi Protected Access (WPA, WPA2, WPA3), to secure wireless computer networks. Many access points will also offer Wi-Fi Protected Setup, a quick (but now insecure) method of joining a new device to an encrypted network.

Infrastructure

Most Wi-Fi networks are deployed in infrastructure mode.

In infrastructure mode, a base station acts as a wireless access point hub, and nodes communicate through the hub. The hub usually, but not always, has a wired or fiber network connection, and may have permanent wireless connections to other nodes.

Wireless access points are usually fixed, and provide service to their client nodes within range.

Wireless clients, such as laptops, smartphones etc. connect to the access point to join the network.

Sometimes a network will have a multiple access points, with the same 'SSID' and security arrangement. In that case connecting to any access point on that network joins the client to the network. In that case, the client software will try to choose the access point to try to give the best service, such as the access point with the strongest signal.

Peer-to-Peer

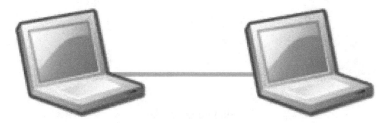

Peer-to-Peer or ad hoc wireless LAN

An ad hoc network (not the same as a WiFi Direct network) is a network where stations communicate only peer to peer (P2P). There is no base and no one gives permission to talk. This is accomplished using the Independent Basic Service Set (IBSS).

A WiFi Direct network is another type of network where stations communicate peer to peer.

In a Wi-Fi P2P group, the group owner operates as an access point and all other devices are clients. There are two main methods to establish a group owner in the Wi-Fi Direct group. In one approach, the user sets up a P2P group owner manually. This method is also known as Autonomous Group Owner (autonomous GO). In the second method, also called negotiation-based group creation, two devices compete based on the group owner intent value. The device with higher intent value becomes a group owner and the second device becomes a client. Group owner intent value can depend on whether the wireless device performs a cross-connection between an infrastructure WLAN service and a P2P group, remaining power in the wireless device, whether the wireless device is already a group owner in another group and/or a received signal strength of the first wireless device.

A peer-to-peer network allows wireless devices to directly communicate with each other. Wireless devices within range of each other can discover and communicate directly without involving central access points. This method is typically used by two computers so that they can connect to each other to form a network. This can basically occur in devices within a closed range.

If a signal strength meter is used in this situation, it may not read the strength accurately and can be misleading, because it registers the strength of the strongest signal, which may be the closest computer.

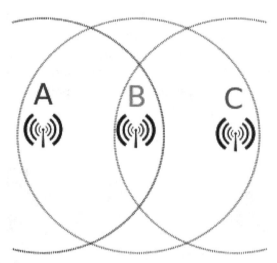

Hidden node problem:
Devices A and C are both communicating with B, but are unaware of each other

IEEE 802.11 defines the physical layer (PHY) and MAC (Media Access Control) layers based on CSMA/CA (Carrier Sense Multiple Access with Collision Avoidance). This is in contrast to Ethernet which uses CSMA-CD (Carrier Sense Multiple Access with Collision Detection). The 802.11 specification includes provisions designed to minimize collisions, because two mobile units may both be in range of a common access point, but out of range of each other.

Bridge

A bridge can be used to connect networks, typically of different types. A wireless Ethernet bridge allows the connection of devices on a wired Ethernet network to a wireless network. The bridge acts as the connection point to the Wireless LAN.

Wireless Distribution System

A Wireless Distribution System enables the wireless interconnection of access points in an IEEE 802.11 network. It allows a wireless network to be expanded using multiple access points without the need for a wired backbone to link them, as is traditionally required. The notable advantage of DS over other solutions is that it preserves the MAC addresses of client packets across links between access points.

An access point can be either a main, relay or remote base station. A main base station is typically connected to the wired Ethernet. A relay base station relays data between remote base stations, wireless clients or other relay stations to either a main or another relay base station. A remote base station accepts connections from wireless clients and passes them to relay or main stations. Connections between "clients" are made using MAC addresses rather than by specifying IP assignments.

All base stations in a Wireless Distribution System must be configured to use the same radio channel, and share WEP keys or WPA keys if they are used. They can be configured to different service set identifiers. WDS also requires that every base station be configured to forward to others in the system as mentioned above.

WDS may also be referred to as repeater mode because it appears to bridge and accept wireless clients at the same time (unlike traditional bridging). Throughput in this method is halved for all clients connected wirelessly.

When it is difficult to connect all of the access points in a network by wires, it is also possible to put up access points as repeaters.

Roaming

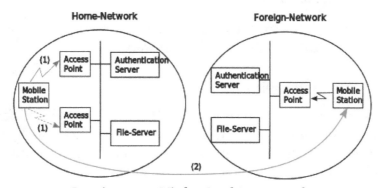

Roaming among Wireless Local Area Networks

There are two definitions for wireless LAN roaming:

1. Internal roaming: The Mobile Station (MS) moves from one access point (AP) to another AP within a home network if the signal strength is too weak. An authentication server (RADIUS) performs the re-authentication of MS via 802.1x (e.g. with PEAP). The billing of QoS is in the home network. A Mobile Station roaming from one access point to another often interrupts the flow of data among the Mobile Station and an application connected to the network. The Mobile Station, for instance, periodically monitors the presence of alternative access points (ones that will provide a better connection). At some point, based on proprietary mechanisms, the Mobile Station decides to re-associate with an access point having a stronger wireless signal. The Mobile Station, however, may lose a connection with an access point before associating with another access point. In order to provide reliable connections with applications, the Mobile Station must generally include software that provides session persistence.

2. External roaming: The MS (client) moves into a WLAN of another Wireless Internet Service Provider (WISP) and takes their services (Hotspot). The user can use a foreign network independently from their home network, provided that the foreign network allows visiting users on their network. There must be special authentication and billing systems for mobile services in a foreign network.

Applications

Wireless LANs have a great deal of applications. Modern implementations of WLANs range from small in-home networks to large, campus-sized ones to completely mobile networks on airplanes and trains.

Users can access the Internet from WLAN hotspots in restaurants, hotels, and now with portable devices that connect to 3G or 4G networks. Oftentimes these types of public access points require no registration or password to join the network. Others can be accessed once registration has occurred and/or a fee is paid.

Existing Wireless LAN infrastructures can also be used to work as indoor positioning systems with no modification to the existing hardware.

Performance and Throughput

WLAN, organised in various layer 2 variants (IEEE 802.11), has different characteristics. Across all flavours of 802.11, maximum achievable throughputs are either given based on measurements under ideal conditions or in the layer 2 data rates. This, however, does not apply to typical deployments in which data are being transferred between two endpoints of which at least one is typically connected to a wired infrastructure and the other endpoint is connected to an infrastructure via a wire-less link.

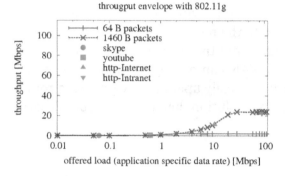

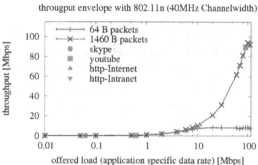

Graphical representation of Wi-Fi application spe-cific (UDP) performance envelope 2.4 GHz band, with 802.11g

Graphical representation of Wi-Fi application spe-cific (UDP) performance envelope 2.4 GHz band, with 802.11n with 40MHz

This means that typically data frames pass an 802.11 (WLAN) medium and are being converted to 802.3 (Ethernet) or vice versa.

Due to the difference in the frame (header) lengths of these two media, the packet size of an application determines the speed of the data transfer. This means that an appli-cation which uses small packets (e.g. VoIP) creates a data flow with a high overhead traffic (e.g. a low goodput).

Other factors which contribute to the overall application data rate are the speed with which the application transmits the packets (i.e. the data rate) and the energy with which the wireless signal is received.

The latter is determined by distance and by the configured output power of the com-municating devices.

Same references apply to the attached throughput graphs which show measurements of UDP throughput measurements. Each represents an average (UDP) throughput (the error bars are there, but barely visible due to the small variation) of 25 measure-ments.

Each is with a specific packet size (small or large) and with a specific data rate (10 kbit/s – 100 Mbit/s). Markers for traffic profiles of common applications are included as well. This text and measurements do not cover packet errors but information about this can be found at above references. The graph above shows the maximum achievable (appli-cation specific) UDP throughput in the same scenarios (same references again) with various difference WLAN (802.11) flavours. The measurement hosts have been 25 me-ters apart from each other; loss is again ignored.

Advantages of WLANs

The most obvious advantage of a WLAN is that devices can connect wirelessly, eliminating the need for cables. This allows homes and businesses to create local networks without wiring the building with Ethernet. It also provides a way for small devices, such as smartphones and tablets, to connect to the network. WLANs are not limited by the number of physical ports on the router and therefore can support dozens or even hundreds of devices. The range of a WLAN can easily be extended by adding one or more repeaters. Finally, a WLAN can be easily upgraded by replacing routers with new versions — a much easier and cheaper solution than upgrading old Ethernet cables.

Disadvantages of WLANs

Wireless networks are naturally less secure than wired networks. Any wireless device can attempt to connect to a WLAN, so it is important to limit access to the network if security is a concern. This is typically done using wireless authentication such as WEP or WPA, which encrypts the communication. Additionally, wireless networks are more susceptible to interference from other signals or physical barriers, such as concrete walls. Since LANs offer the highest performance and security, they are still used for many corporate and government networks.

Wireless WAN

A wireless wide area network (WWAN) is a specific type of network that sends wireless signals beyond a single building or property. By contrast, a local area network or LAN connects computers and other hardware pieces inside a residential or commercial property. Wireless wide area networks and wireless local area networks also differ in the types of signal processing technologies they use.

While local area networks often rely on Ethernet, twisted-pair cabling or short-range wireless routers, a wireless WAN may use various types of cellular network systems to send signals over a longer distance. Large telecom providers like T-Mobile, Sprint, Verizon and AT&T typically support a wireless WAN in one way or another, and these larger types of networks often require some types of encryption or security that a local area network may not need.

Because wireless wide area networks rely on the same telecom systems that support the delivery of data and voice to and from modern tablet and smartphone devices, these larger types of networks may also be vulnerable to what's called a spectrum crunch, where a current shortage in the limited amount of wireless spectrum frequencies may have an impact on how telecom providers can deliver services to a growing consumer base. This may cause some wireless wide area network administrators to change elements of their networks in order to rely less on systems that are reaching a point of maximum capacity.

The three major wireless WAN technologies comprise the two traditional cellular systems, GSM and CDMA, and the newer WiMAX. GSM and CDMA use HSPA and EV-DO to deliver 3G data rates. WiMAX delivers faster data service.

Short Message Service (SMS) Applications

One of the most common services for wireless WANs is short message service (SMS), which is a text messaging system capable of sending a couple hundred characters at a time. SMS is a wireless form of the familiar instant messaging that is available from many of the ISPs. The following are additional applications of SMS for use with wireless WANs:

- Content delivery? SMS enables the efficient delivery of updates for user devices. For example, a user can download new ring tones and fancy backgrounds to her phone over SMS. In addition, SMS makes it possible for users to query databases and news feeds. For example, you can keep up with the latest breaking news by receiving instant updates through SMS.

- Alerts? Most operators offer a variety of alerts, such as voicemail waiting, sports scores, current stock prices, and other reminders. This allows you to receive updates when something that you define as important occurs.

- Interaction? Some television shows allow interactivity among viewers and hosts through SMS. This enables participation of the audience, which dramatically increases the viewing audience size

- Application integration? It's feasible for developers to integrate SMS into many corporate applications. For example, a company can have a sales management system that enables sales representatives to track customers and products. A company can usually easily add SMS to these applications. For example, the addition of an alert mechanism would be beneficial. Sales representatives would be notified that a product has gone on sale.

Many web sites use Wireless Markup Language (WML) to transform regular web pages into a format that is more easily read on a small device, such as a PDA or cell phone. WML also reduces the graphics on the page to compensate for the slower data rates of wireless WAN technologies.

A WWAN may also be a closed network that covers a large geographic area. For example, a mesh network or MANET with nodes on building, tower, trucks, and planes could also be considered a WWAN.

A WWAN is also different from a low-power, low-bit-rate wireless WAN, (LPWAN), intended to carry small packets of information between things, often in the form of battery operated sensors.

Since radio communications systems do not provide a physically secure connection path, WWANs typically incorporate encryption and authentication methods to make them more secure. Unfortunately some of the early GSM encryption techniques were flawed, and security experts have issued warnings that cellular communication, including WWAN, is no longer secure. UMTS (3G) encryption was developed later and has yet to be broken.

Wireless PAN

A *wireless personal area network* (WPAN) is a personal, short distance area wireless network for interconnecting devices centered around an individual person's workspace. WPANs address wireless networking and mobile computing devices such as PCs, PDAs, peripherals, cell phones, pagers and consumer electronics. WPANs are also called *short wireless distance networks*.

Typically, a wireless personal area network uses some technology that permits communication within about 10 meters - in other words, a very short range. One such technology is Bluetooth, which was used as the basis for a new standard, IEEE 802.15.

A WPAN could serve to interconnect all the ordinary computing and communicating devices that many people have on their desk or carry with them today - or it could serve a more specialized purpose such as allowing the surgeon and other team members to communicate during an operation.

A key concept in WPAN technology is known as *plugging in*. In the ideal scenario, when any two WPAN-equipped devices come into close proximity (within several meters of each other) or within a few kilometers of a central server, they can communicate as if connected by a cable. Another important feature is the ability of each device to lock out other devices selectively, preventing needless interference or unauthorized access to information.

The technology for WPANs is in its infancy and is undergoing rapid development. Proposed operating frequencies are around 2.4 GHz in digital modes. The objective is to facilitate seamless operation among home or business devices and systems. Every device in a WPAN will be able to plug in to any other device in the same WPAN, provided they are within physical range of one another. In addition, WPANs worldwide will be interconnected. Thus, for example, an archeologist on site in Greece might use a PDA to directly access databases at the University of Minnesota in Minneapolis, and to transmit findings to that database.

IEEE 802.15 has produced standards for several types of PANs operating in the ISM band including Bluetooth. The Infrared Data Association has produced standards for WPANs which operate using infrared communications.

A WPAN could serve to interconnect all the ordinary computing and communicating devices that many people have on their desk or carry with them today; or it could serve a more specialized purpose such as allowing the surgeon and other team members to communicate during an operation.

A key concept in WPAN technology is known as *plugging in*. In the ideal scenario, when any two WPAN-equipped devices come into close proximity (within several meters of each other) or within a few kilometers of a central server, they can communicate as if connected by a cable. Another important feature is the ability of each device to lock out other devices selectively, preventing needless interference or unauthorized access to information.

The technology for WPANs is in its infancy and is undergoing rapid development. Proposed operating frequencies are around 2.4 GHz in digital modes. The objective is to facilitate seamless operation among home or business devices and systems. Every device in a WPAN will be able to plug into any other device in the same WPAN, provided they are within physical range of one another. In addition, WPANs worldwide will be interconnected. Thus, for example, an archeologist on site in Greece might use a PDA to directly access databases at the University of Minnesota in Minneapolis, and to transmit findings to that database.

Bluetooth

Bluetooth uses short-range radio waves. While historically covering shorter distances associated with a PAN, the Bluetooth 5 standard, Bluetooth Mesh, have extended that range considerably. Further, long range Bluetooth routers with augmented antenna arrays connect Bluetooth devices up to 1,000 feet. Uses in a PAN include, for example, Bluetooth devices such as keyboards, pointing devices, audio head sets, printers may connect to personal digital assistants (PDAs), cell phones, or computers.

A Bluetooth PAN is also called a *piconet* (combination of the prefix "pico," meaning very small or one trillionth, and network), and is composed of up to 8 active devices in a master-slave relationship (a very large number of devices can be connected in "parked" mode). The first Bluetooth device in the piconet is the master, and all other devices are slaves that communicate with the master. A piconet typically has a range of 10 metres (33 ft), although ranges of up to 100 metres (330 ft) can be reached under ideal circumstances. With Bluetooth mesh networking the range and number of devices is extended by relaying information from one to another. Such a network doesn't have a master device and may or may not be treated as a PAN.

Infrared Data Association

Infrared Data Association (IrDA) uses infrared light, which has a frequency below the human eye's sensitivity. Infrared in general is used, for instance, in TV remotes. Typical WPAN devices that use IrDA include printers, keyboards, and other serial communication interfaces.

WPAN Comparison

Of the three WPAN standards, IrDA, Bluetooth, and 802.15, IrDA has been around the longest, and has the highest market penetration, with more than 300 million enabled devices shipped. At the same time, infrared also is the most limiting, as the range is up to 2 meters, and it requires a line of site between communicating devices. The Bluetooth specification addresses these issues by using unlicensed 2.4-GHz spectrum for communication. This allows for communication through physical barriers, as well as larger ranges, typically up to about 10 meters. Bluetooth has also garnered a lot of industry attention, with more than 2,000 companies joining the Bluetooth SIG. In order to provide further standardization for WPAN technology, the IEEE 802.15 specification was developed. The 802.15 specification uses Bluetooth v1.1 as a foundation for providing standardized short-range wireless communication between portable and mobile computing devices. Table below provides a summary of the leading WPAN technologies.

Table: Comparison of WPAN Technologies				
Standard	Frequency	Bandwidth	Optimum Operating Range	Points Of Interest
IrDA	875nm wavelength	9600 bps to 4 Mbps. Future of 15 Mbps	1-2 meters (3–6 feet)	Requires line of site for communication.
Bluetooth	2.4 GHz	v1.1: 720 Kbps; v2.0: 10 Mbps	10 meters (30 feet) to 100 meters (300 feet)	Automatic device discovery; communicates through physical barriers.
IEEE 802.15	2.4 GHz	802.15.1: 1 Mbps 802.15.3: 20-plus Mbps	10 meters (30 feet) to 100 meters (300 feet)	Uses Bluetooth as the foundation; coexistence with 802.11 devices.

There is no clear leader, as we are still in the early stages of WPAN technology development. Bluetooth has generated the most industry attention so far, but 802.15 is just as exciting. Since 802.15 is interoperable with both Bluetooth and 802.11, it will have a solid future in the WPAN space. In many ways, IR is not a competing technology to either Bluetooth or 802.15 since it addresses a separate market need. IrDA is included in nearly all mobile devices, providing a quick and easy way for reliable short-range data transfer. With its low implementation costs, many low-end devices will continue to support IrDA, while more advanced devices with more robust wireless needs will implement Bluetooth or 802.15.

Another area of interest is the increasing range that these technologies can address. Initially, Bluetooth was aimed at a personal operating space of 10 meters. Now, with power-amplified Bluetooth access points, the range has extended to 100 meters. 802.15 is in the same situation. The increased range for these technologies blurs the line between wireless personal area networks and wireless local area networks.

Wireless MAN

Wireless metropolitan area networks (WMANs) enable users to establish wireless connections between multiple locations within a metropolitan area (for example, between multiple office buildings in a city or on a university campus), without the high cost of laying fiber or copper cabling and leasing lines. In addition, WMANs can serve as backups for wired networks, should the primary leased lines for wired networks become unavailable. WMANs use either radio waves or infrared light to transmit data. Broadband wireless access networks, which provide users with high-speed access to the Internet, are in increasing demand.

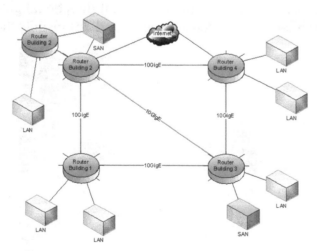

Wireless MAN Systems

Wireless MANs offer connections between buildings and users within a city or campus area through several system configurations. In most cases, the wireless MAN beams RF or infrared light from one point to another using directive antennae.

Point-to-Point Systems

A point-to-point solution uses RF or infrared signals that utilize either semidirectional or highly directional antennae to extend range across metropolitan areas, such as college campuses and cities. Range can be as high as 30 miles for RF systems using highly directional antennae. Figure below illustrates a point-to-point wireless MAN system.

Figure- Point-to-Point Wireless MAN Directly Connects Two Points in the Network

A medical center, for example, can use a point-to-point wireless MAN to provide a communications link between the main hospital and a remote clinic within the same city. This resulting system, however, does not provide as much flexibility as point-to-multipoint solutions. However, if there is a need to connect only a couple sites, the cost of implementing a point-to-point system is less compared to a point-to-multipoint system.

Point-to-Multipoint System

A typical point-to-multipoint link utilizes a centralized omnidirectional antenna that provides a single transceiver point for tying together multiple remote stations. For example, a building within the center of a city can host the omnidirectional antenna, and other nearby metropolitan-area buildings can point directional antennae at the centralized location. The central transceiver receives and retransmits the signals.

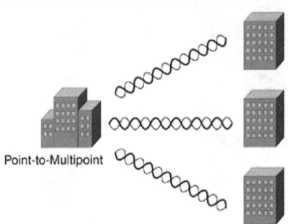

Figure- Point-to-Multipoint Wireless MAN Interconnects
Users Through a Common, Centralized Transceiver

A strong advantage of the point-to-multipoint wireless MAN is that it makes the addition of new connections easy. In fact, this approach can be less expensive compared to point-to-point systems when there are multiple sites to interconnect or connect to a central location. For example, a company headquarters having many remote warehouses and manufacturing plants within the same city or rural area would benefit from a point-to-multipoint system.

Packet Radio Systems

A packet radio system utilizes special wireless routers that forward data contained within packets to the destination. Each user has a packet radio NIC that transmits data to the nearest wireless router. This router then retransmits the data to the next router. This hopping from router to router occurs until the packet reaches the destination. This mesh type networking is not new. Amateur Ham radio operators have used it for decades, and companies such as Metricom have been deploying these types of systems in cities for nearly 10 years.

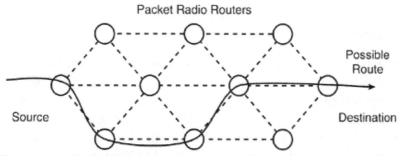

Figure- Packet Radio System Hops Data Packets from the Source to Destination

A city government might want to deploy a packet radio system to offer wireless connectivity for supporting applications through the entire city area. The installation of routers in strategic places through the city provides the necessary infrastructure. There's no need for wires for interconnecting the routers. Each router is capable of receiving and retransmitting? hopping? the packets to their destination.

This form of networking is also survivable. If one router becomes inoperative, perhaps because of a lightning strike or sabotage, adaptive routing protocols automatically update routing tables in each router so that data packets will avoid traversing the inoperative router.

WMANs Standards

There are currently three different standards being researched and produced to support WMANs:

- HiperMAN

- HiperACCESS

- 802.16

HiperMAN & HiperACCESS

High Performance Radio Access (HIPERACCESS) is an interoperable standard tailored to give broadband access to both the home and small- and medium-sized enterprises, as well as to provide backhaul for mobile systems (e.g. W-CDMA, CDMA2000, GSM and GPRS). The ETSI Project Broadband Radio Access Networks (EP BRAN) working on the standard agreed the core technical specifications (i.e. the specifications for the Physical (PHY) Layer and for the Data Link Control (DLC) Layer) in early 2002, and is in the process of developing the convergence layer for support of different core networks, such as Internet Protocol (IP), Ethernet and Asynchronous Transfer Mode (ATM).

The HIPERACCESS standard was developed to provide a truly broadband system with bit rates of up to approximately 100 Mbit/s, although 25Mbit/sec is expected to be the most widely deployed rate. HIPERACCESS is targeted at high frequency bands, especially for the 40,5 - 43,5 GHz band. For these frequency bands TDMA will be used to provide multiple access.

High Performance Radio Metropolitan Area Network (HIPERMAN) is aimed at providing a broadband wireless solution for Metropolitan Area Networks. HIPER-MAN will be an interoperable broadband fixed wireless access system operating at radio frequencies between 2 GHz and 11 GHz. The air interface will be optimized for Point-to-MultiPoint (PMP) configurations, but allows for flexible mesh network deployments as well. HIPERACCESS and HIPERMAN address the same markets but have been separated as the higher frequencies used in HIPERACCESS require different techniques to the lower frequencies used by HIPERMAN to deliver the required QoS and other system metrics.

The IEEE 802.16 WirelessMAN Standard

The Institute of Electrical and Electronics Engineers Standards Association (IEEE-SA) sought to make Broadband Wireless Access (BWA) more widely available by developing IEEE Standard 802.16, which specifies the WirelessMAN Air Interface for wireless metropolitan area networks. The standard, which was published on 8 April 2002, was created in a two-year, open-consensus process by hundreds of engineers from the world's leading operators and vendors.

IEEE 802.16 addresses the "first-mile/last-mile" connection in wireless metropolitan area networks. It focuses on the efficient use of bandwidth between 10 and 66 GHz (the 2 to 11 GHz region with PMP and optional Mesh topologies by the end of 2002) and defines a medium access control (MAC) layer that supports multiple physical layer specifications customized for the frequency band of use.

The 10 to 66 GHz standard supports continuously varying traffic levels at many licensed frequencies (e.g., 10.5, 25, 26, 31, 38 and 39 GHz) for two-way communications. It enables interoperability among devices, so carriers can use products from multiple vendors and warrants the availability of lower cost equipment. The draft amendment for the 2 to 11 GHz region will support both unlicensed and licensed bands.

Types of WMANs

There are two basic types of WMANs: back haul and last mile.

Back Haul

Back haul is for enterprise networks, cellular-tower connection and Wi-Fi hotspots. It's an option for enterprises that can't afford to install or lease fiber to connect their facilities over a large campus or city. Back-haul WMANs also make sense when you can't justify a service provider's fiber 10-Mbps connection that requires six T1 leased lines at $50,000 a year. Fixed wireless is about half that price, without a monthly charge.

Although DSL or T1 work for back haul, a private broadband wireless system often provides 10 times faster transmissions. Capital and installation costs are about five times higher, but you get an ROI in just a few months. To interconnect a few sites, you can install one or more wireless PTP (point-to-point) links; for more sites, a private multipoint WMAN usually makes more sense.

Last Mile

Wireless back haul has the greatest short-term appeal, but last-mile solutions could establish wireless as an alternative to residential broadband DSL/cable modem. Some wireless ISPs, including TowerStream Corp., compete head-on with broadband, offering quick installation and lower cost, as well as wireless Internet service in areas without access. Last-mile WMANs are handy for temporary networks, too, such as large construction sites or areas where conventional network service is disrupted.

WMAN Models

There are several models for metropolitan area networks. The simplest and most common (and not metro-wide at all) is the downtown hotzone. A hotzone is a contiguous cluster of Wi-Fi hotspots. Most cities set up downtown hotzones to promote themselves as business and high-tech centers. In some cities, public agencies, companies, hospitals, universities and community groups are knitting together hotspots into a network to provide broadband service to their members. The ambitions of these city projects are modest, but if they become popular, they encourage cities to expand the public access wireless network.

A few cities deploy a wireless network over a larger area, for example, a disadvantaged neighborhood. The purpose is to provide broadband service to a people whom the cable and DSL providers have abandoned. An example is Manchester 's EastServe which is deployed over a large neighborhood composed primarily of council houses and people on welfare. The ambitions of cities in building out these types of wireless broadband networks are modest and limited in scope. However, if they prove to be successful in revitalizing entire communities, they give cities another reason to invest in additional wireless broadband infrastructure.

The most ambitious type of wireless metropolitan area network is one which covers an entire city. There are already several successful city-wide networks operating in Finland, albeit in very small cities such as Vantaa and Porvoo. Truly large-scale wireless metropolitan area networks over mega-cities such as New York or Paris do not exist. There are many reasons for this. Cities do not see why they, instead of private parties, should build the network. Moreover, the costs are astronomical. Nevertheless, with the launch of several hardware products that make metro-wide wireless networks much cheaper and easier to deploy, some cities, such as New York, see the deployment of a wireless broadband network as a means for lowering the city government's massive telecommunications costs and fulfilling social/economic goals.

Advantages of WMANs

The advantages of using a WMAN are simple enough to distinguish. Intially, the cost of a WMAN is significantly less then fiber-based LANs for a number of reasons. WMANs extend, replaces or backs up existing fiber infrastructre within hours, saving time and money on personnel. WMANs also eliminates fiber trenching and leased line costs, a large portion of the cost in laying fiber.

Another significant advantage of implementing a WMAN is the fact that it bridges the "digital divide" by affordingly connecting communities that have not had the access to LANs and wireless internet connections.

Finally, the installation of a WMAN empowers companies, government agencies, etc. to rapidly expand their internet service into new markets and generate increased revenues.

Disadvantages of WMANs

One main disadvantage of a WMAN presents a very large area for a hacker to attempt a break in. Like any wireless access network, a wireless access point is a tempting target for someone to hack into a secure network. A WMAN is simply a larger WAN footprint, and thus presents a larger opportunity for a hacker.

The Producers and Current users of this Technology

Below are descriptions of the different types of wireless networks deployed in various cities.

Examples of downtown hotzones

- Hamburg : Germany 's biggest non-commercial WLAN. It is the centerpiece of the Hamburg -Always On program, which is run by Hamburg@work, Deutsche

Telekom, Fujitsu Siemens Computers, Datenlotsen Information Systems and Siggelkow Computers. Hamburg@work is the city's initiative for new media, IT and telecommunications. According to the Hamburg@work, it is the most comprehensive free WLAN at the municipal level. There are about 41 hotspots with more planned for 2003 and 2004. Recently, Hotspot Hamburg added the luxurious Hotel Vier Jahreszeiten to its network. Guests and visitors enjoy free wireless access to the Internet in all of the hotel's rooms, in the lobby, convention areas, and the Indochine Bar. How does Hotspot Hamburg work? Locations such as restaurants, cafes, libraries and hotels receive the necessary hardware and software for free from the sponsors. The locations pay the ongoing volume-based charges for the T-DSL line which start at 50 Euros per month including 5 gigabytes of data volume. Each location has a special HOTSPOT HAMBURG logo indicating its participation in the network. The sponsors finance the hardware, installation, service and support for the individual WLANs and in return receive exclusive advertising space in the welcome screens, marketing materials and online sites. Travelers often look for Hotspot Hamburg locations signs first before going to locations that make them pay for access.

- Bochum : The Bochum downtown hotzone initiative is called "Kabel ab" (without cable). The network went live in May 2003 and boasts fifty hotspots in the Bermuda Dreieck (Bermuda Triangle) area which has a lot of cafés and beer gardens. Unlike Hamburg Hotspot, the network plans to charge for Wi-Fi access in the future, although for now, it is free. There are no connect time or data volume limits, but you do have to get a pre-paid scratch card (although this is still for free) with a personal code to use it. Bochum will add railway stations and the airport to the network soon. Kabel Ab is a joint venture among the City Council, TMR (local telco operator funded by local electricity companies and savings bank), Intel and the Bochum tourist information service (Bochum Marketing).

- Skellefteå (Sweden): Residents have access to wireless broadband connections at the city center, the campsite and sports complex. Next in line are the airport, hotels, golf course and other facilities. This is a project of Mobile City , conceived two years ago in Sweden , which is dedicated to research, testing and prototype development in wireless communications. The project has received financing from the EU structural funds, the County Council of Västerbotten and Skellefteå Municipality .

- Hasselt : city-wide public wireless access, according to Sinfilo, a Belgian wireless LAN operator. Sinfilo will be building out the network for the city and maintaining it as well. The first phase of the project consists of creating a wireless network in the city's main square, the Grote Markt. The later phases will expand wireless access throughout the city.

- Wellington (New Zealand): 100 hotspots and by November this year, the Wi-Fi service called CafeNet, will be available as well in Lambton Quay, the train station and other busy locations. CafeNet was set up by CityLink, a fibre optic network operator that was originally established by the Wellington City Council in 1995, but is today privately owned (although it has received recently $95,000 from the City Council to expand the CafeNet service). Like other municipalities setting up hotzones or building an entire wireless broadband infrastructure throughout the entire city, Wellington 's purpose in supporting CityLink is to promote economic development and make it more attractive for technology industries and workers.

- Adelaide (Australia): Australia 's largest public access wireless network is in its central business district and north Adelaide with 40 access points, each with a range of 200 meters line of sight. One of the reasons for setting up the network is to boost the state's IT reputation, and provide more opportunities for local startup companies. Many of these hotspots use city infrastructure such as traffic lights and lamp posts, identifiable by the citilan decal. They claim that anyone using a laptop or personal digital assistant (PDA) enabled with a wireless card within range of these emblems will be able to access citilan and roam across the city while still connected. The network is the result of a partnership among the Adelaide City Council, service providers Internode/Agile and AirNet, Cisco, the South Australian Government and mobile broadband consortium m.Net Corporation.

Other cities that launched hotzones (recently reported on Muniwireless.com) are: Bellevue (Washington state), Milwaukee , Cleveland , St. Louis , Baton Rouge (Louisiana), Portland (Oregon), Baltimore and Pittsburgh .

Cellular Network

A cellular network is a radio network distributed over land through cells where each cell includes a fixed location transceiver known as base station. These cells together provide radio coverage over larger geographical areas. User equipment (UE), such as mobile phones, is therefore able to communicate even if the equipment is moving through cells during transmission.

Cellular networks give subscribers advanced features over alternative solutions, including increased capacity, small battery power usage, a larger geographical coverage area and reduced interference from other signals. Popular cellular technologies include the Global System for Mobile Communication, general packet radio service, 3GSM and code division multiple access.

Cellular network technology supports a hierarchical structure formed by the base transceiver station (BTS), mobile switching center (MSC), location registers and public switched telephone network (PSTN). The BTS enables cellular devices to make direct communication with mobile phones. The unit acts as a base station to route calls to the destination base center controller. The base station controller (BSC) coordinates with the MSC to interface with the landline-based PSTN, visitor location register (VLR), and home location register (HLR) to route the calls toward different base center controllers.

Cellular networks maintain information for tracking the location of their subscribers' mobile devices. In response, cellular devices are also equipped with the details of appropriate channels for signals from the cellular network systems. These channels are categorized into two fields:

- Strong Dedicated Control Channel: Used to transmit digital information to a cellular mobile phone from the base station and vice versa.

- Strong Paging Channel: Used for tracking the mobile phone by MSC when a call is routed to it.

A typical cell site offers geographical coverage of between nine and 21 miles. The base station is responsible for monitoring the level of the signals when a call is made from a mobile phone. When the user moves away from the geographical coverage area of the base station, the signal level may fall. This can cause a base station to make a request to the MSC to transfer the control to another base station that is receiving the strongest signals without notifying the subscriber; this phenomenon is called handover. Cellular networks often encounter environmental interruptions like a moving tower crane, overhead power cables, or the frequencies of other devices.

Features of Cellular Systems

Wireless Cellular Systems solves the problem of spectral congestion and increases user capacity. The features of cellular systems are as follows –

- Offer very high capacity in a limited spectrum.

- Reuse of radio channel in different cells.

- Enable a fixed number of channels to serve an arbitrarily large number of users by reusing the channel throughout the coverage region.

- Communication is always between mobile and base station (not directly between mobiles).

- Each cellular base station is allocated a group of radio channels within a small geographic area called a cell.

- Neighboring cells are assigned different channel groups.

- By limiting the coverage area to within the boundary of the cell, the channel groups may be reused to cover different cells.

- Keep interference levels within tolerable limits.

- Frequency reuse or frequency planning.

- Organization of Wireless Cellular Network.

Cellular network is organized into multiple low power transmitters each 100w or less.

Concept

In a cellular radio system, a land area to be supplied with radio service is divided into cells, in a pattern which depends on terrain and reception characteristics but which can consist of roughly hexagonal, square, circular or some other regular shapes, although hexagonal cells are conventional. Each of these cells is assigned with multiple frequencies $(f_1 - f_6)$ which have corresponding radio base stations. The group of frequencies can be reused in other cells, provided that the same frequencies are not reused in adjacent neighboring cells as that would cause co-channel interference.

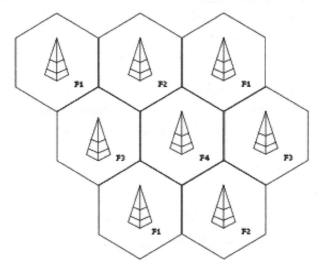

Example of frequency reuse factor or pattern 1/4

The increased capacity in a cellular network, compared with a network with a single transmitter, comes from the mobile communication switching system developed by Amos Joel of Bell Labs that permitted multiple callers in the same area to use the same frequency by switching calls made using the same frequency to the nearest available cellular tower having that frequency available and from the fact that the same radio frequency can be reused in a different area for a completely different transmission. If there is a single plain transmitter, only one transmission can be used on any given frequency. Unfortunately, there is inevitably some level of interference from the signal from the other cells which use the same frequency. This means that, in a standard FDMA system, there must be at least a one cell gap between cells which reuse the same frequency.

In the simple case of the taxi company, each radio had a manually operated channel selector knob to tune to different frequencies. As the drivers moved around, they would change from channel to channel. The drivers knew which frequency covered approximately what area. When they did not receive a signal from the transmitter, they would try other channels until they found one that worked. The taxi drivers would only speak one at a time, when invited by the base station operator. This is, in a sense, time-division multiple access (TDMA).

The first commercially cellular network, the 1G generation, was launched in Japan by Nippon Telegraph and Telephone (NTT) in 1979, initially in the metropolitan area of Tokyo. Within five years, the NTT network had been expanded to cover the whole population of Japan and became the first nationwide 1G network.

Cell Signal Encoding

To distinguish signals from several different transmitters, time-division multiple access (TDMA), frequency-division multiple access (FDMA), code-division multiple access (CDMA), and orthogonal frequency-division multiple access (OFDMA) were developed.

With TDMA, the transmitting and receiving time slots used by different users in each cell are different from each other.

With FDMA, the transmitting and receiving frequencies used by different users in each cell are different from each other. In a simple taxi system, the taxi driver manually tuned to a frequency of a chosen cell to obtain a strong signal and to avoid interference from signals from other cells.

The principle of CDMA is more complex, but achieves the same result; the distributed transceivers can select one cell and listen to it.

Other available methods of multiplexing such as polarization-division multiple access (PDMA) cannot be used to separate signals from one cell to the next since the effects of both vary with position and this would make signal separation practically impossible. TDMA is used in combination with either FDMA or CDMA in a number of systems to give multiple channels within the coverage area of a single cell.

Frequency Reuse

The key characteristic of a cellular network is the ability to re-use frequencies to increase both coverage and capacity. As described above, adjacent cells must use different frequencies, however there is no problem with two cells sufficiently far apart operating on the same frequency, provided the masts and cellular network users' equipment do not transmit with too much power.

The elements that determine frequency reuse are the reuse distance and the reuse factor. The reuse distance, D is calculated as

$$D = R\sqrt{3N}$$,

where R is the cell radius and N is the number of cells per cluster. Cells may vary in radius from 1 to 30 kilometres (0.62 to 18.64 mi). The boundaries of the cells can also overlap between adjacent cells and large cells can be divided into smaller cells.

The frequency reuse factor is the rate at which the same frequency can be used in the network. It is $1/K$ (or K according to some books) where K is the number of cells which cannot use the same frequencies for transmission. Common values for the frequency reuse factor are 1/3, 1/4, 1/7, 1/9 and 1/12 (or 3, 4, 7, 9 and 12 depending on notation).

In case of N sector antennas on the same base station site, each with different direction, the base station site can serve N different sectors. N is typically 3. A reuse pattern of N/K denotes a further division in frequency among N sector antennas per site. Some current and historical reuse patterns are 3/7 (North American AMPS), 6/4 (Motorola NAMPS), and 3/4 (GSM).

If the total available bandwidth is B, each cell can only use a number of frequency channels corresponding to a bandwidth of B/K, and each sector can use a bandwidth of B/NK.

Code-division multiple access-based systems use a wider frequency band to achieve the same rate of transmission as FDMA, but this is compensated for by the ability to use a frequency reuse factor of 1, for example using a reuse pattern of 1/1. In other words, adjacent base station sites use the same frequencies, and the different base stations and users are separated by codes rather than frequencies. While N is shown as 1 in this example, that does not mean the CDMA cell has only one sector, but rather that the entire cell bandwidth is also available to each sector individually.

Depending on the size of the city, a taxi system may not have any frequency-reuse in its own city, but certainly in other nearby cities, the same frequency can be used. In a large city, on the other hand, frequency-reuse could certainly be in use.

Recently also orthogonal frequency-division multiple access based systems such as LTE are being deployed with a frequency reuse of 1. Since such systems do not spread the signal across the frequency band, inter-cell radio resource management is important to coordinate resource allocation between different cell sites and to limit the inter-cell interference. There are various means of Inter-Cell Interference Coordination (ICIC) already defined in the standard. Coordinated scheduling, multi-site MIMO or multi-site beam forming are other examples for inter-cell radio resource management that might be standardized in the future.

Directional Antennas

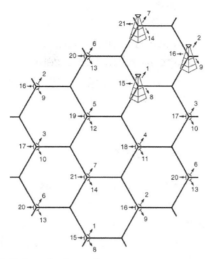

Cellular telephone frequency reuse pattern

Cell towers frequently use a directional signal to improve reception in higher-traffic areas. In the United States, the FCC limits omnidirectional cell tower signals to 100 watts of power. If the tower has directional antennas, the FCC allows the cell operator to broadcast up to 500 watts of effective radiated power (ERP).

Although the original cell towers created an even, omnidirectional signal, were at the centers of the cells and were omnidirectional, a cellular map can be redrawn with the cellular telephone towers located at the corners of the hexagons where three cells converge. Each tower has three sets of directional antennas aimed in three different directions with 120 degrees for each cell (totaling 360 degrees) and receiving/transmitting into three different cells at different frequencies. This provides a minimum of three channels, and three towers for each cell and greatly increases the chances of receiving a usable signal from at least one direction.

The numbers in the illustration are channel numbers, which repeat every 3 cells. Large cells can be subdivided into smaller cells for high volume areas.

Cell phone companies also use this directional signal to improve reception along highways and inside buildings like stadiums and arenas.

Broadcast Messages and Paging

Practically every cellular system has some kind of broadcast mechanism. This can be used directly for distributing information to multiple mobiles. Commonly, for example in mobile telephony systems, the most important use of broadcast information is to set up channels for one-to-one communication between the mobile transceiver and the base station. This is called paging. The three different paging procedures generally adopted are sequential, parallel and selective paging.

The details of the process of paging vary somewhat from network to network, but normally we know a limited number of cells where the phone is located (this group of cells is called a Location Area in the GSM or UMTS system, or Routing Area if a data packet session is involved; in LTE, cells are grouped into Tracking Areas). Paging takes place by sending the broadcast message to all of those cells. Paging messages can be used for information transfer. This happens in pagers, in CDMA systems for sending SMS messages, and in the UMTS system where it allows for low downlink latency in packet-based connections.

Movement from Cell to Cell and Handing Over

In a primitive taxi system, when the taxi moved away from a first tower and closer to a second tower, the taxi driver manually switched from one frequency to another as needed. If a communication was interrupted due to a loss of a signal, the taxi driver asked the base station operator to repeat the message on a different frequency.

In a cellular system, as the distributed mobile transceivers move from cell to cell during an ongoing continuous communication, switching from one cell frequency to a different cell frequency is done electronically without interruption and without a base station operator or manual switching. This is called the handover or handoff. Typically, a new channel is automatically selected for the mobile unit on the new base station which will serve it. The mobile unit then automatically switches from the current channel to the new channel and communication continues.

The exact details of the mobile system's move from one base station to the other varies considerably from system to system.

Mobile Phone Network

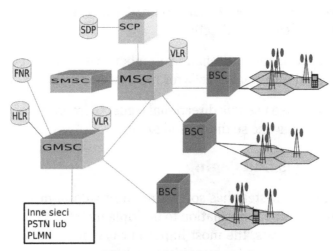

GSM network architecture

The most common example of a cellular network is a mobile phone (cell phone) network.

A mobile phone is a portable telephone which receives or makes calls through a cell site (base station), or transmitting tower. Radio waves are used to transfer signals to and from the cell phone.

Modern mobile phone networks use cells because radio frequencies are a limited, shared resource. Cell-sites and handsets change frequency under computer control and use low power transmitters so that the usually limited number of radio frequencies can be simultaneously used by many callers with less interference.

A cellular network is used by the mobile phone operator to achieve both coverage and capacity for their subscribers. Large geographic areas are split into smaller cells to avoid line-of-sight signal loss and to support a large number of active phones in that area. All of the cell sites are connected to telephone exchanges (or switches), which in turn connect to the public telephone network.

In cities, each cell site may have a range of up to approximately $\frac{1}{2}$ mile (0.80 km), while in rural areas, the range could be as much as 5 miles (8.0 km). It is possible that in clear open areas, a user may receive signals from a cell site 25 miles (40 km) away.

Since almost all mobile phones use cellular technology, including GSM, CDMA, and AMPS (analog), the term "cell phone" is in some regions, notably the US, used interchangeably with "mobile phone". However, satellite phones are mobile phones that do not communicate directly with a ground-based cellular tower, but may do so indirectly by way of a satellite.

There are a number of different digital cellular technologies, including: Global System for Mobile Communications (GSM), General Packet Radio Service (GPRS), cdmaOne, CDMA2000, Evolution-Data Optimized (EV-DO), Enhanced Data Rates for GSM Evolution (EDGE), Universal Mobile Telecommunications System (UMTS), Digital Enhanced Cordless Telecommunications (DECT), Digital AMPS (IS-136/TDMA), and Integrated Digital Enhanced Network (iDEN). The transition from existing analog to the digital standard followed a very different path in Europe and the US. As a consequence, multiple digital standards surfaced in the US, while Europe and many countries converged towards the GSM standard.

Structure of the Mobile Phone Cellular network

A simple view of the cellular mobile-radio network consists of the following:

- A network of radio base stations forming the base station subsystem.
- The core circuit switched network for handling voice calls and text
- A packet switched network for handling mobile data
- The public switched telephone network to connect subscribers to the wider telephony network

This network is the foundation of the GSM system network. There are many functions that are performed by this network in order to make sure customers get the desired service including mobility management, registration, call set-up, and handover.

Any phone connects to the network via an RBS (Radio Base Station) at a corner of the corresponding cell which in turn connects to the Mobile switching center (MSC). The MSC provides a connection to the public switched telephone network (PSTN). The link from a phone to the RBS is called an *uplink* while the other way is termed *downlink*.

Radio channels effectively use the transmission medium through the use of the following multiplexing and access schemes: frequency division multiple access (FDMA), time division multiple access (TDMA), code division multiple access (CDMA), and space division multiple access (SDMA).

Small Cells

Small cells, which have a smaller coverage area than base stations, are categorised as follows:

- Microcell, less than 2 kilometres

- Picocell, less than 200 metres

- Femtocell, around 10 metres

Cellular Handover in Mobile Phone Networks

As the phone user moves from one cell area to another cell while a call is in progress, the mobile station will search for a new channel to attach to in order not to drop the call. Once a new channel is found, the network will command the mobile unit to switch to the new channel and at the same time switch the call onto the new channel.

With CDMA, multiple CDMA handsets share a specific radio channel. The signals are separated by using a pseudonoise code (PN code) specific to each phone. As the user moves from one cell to another, the handset sets up radio links with multiple cell sites (or sectors of the same site) simultaneously. This is known as "soft handoff" because, unlike with traditional cellular technology, there is no one defined point where the phone switches to the new cell.

In IS-95 inter-frequency handovers and older analog systems such as NMT it will typically be impossible to test the target channel directly while communicating. In this case other techniques have to be used such as pilot beacons in IS-95. This means that there is almost always a brief break in the communication while searching for the new channel followed by the risk of an unexpected return to the old channel.

If there is no ongoing communication or the communication can be interrupted, it is possible for the mobile unit to spontaneously move from one cell to another and then notify the base station with the strongest signal.

Cellular Frequency Choice in Mobile Phone Networks

The effect of frequency on cell coverage means that different frequencies serve better for different uses. Low frequencies, such as 450 MHz NMT, serve very well for countryside coverage. GSM 900 (900 MHz) is a suitable solution for light urban coverage. GSM 1800 (1.8 GHz) starts to be limited by structural walls. UMTS, at 2.1 GHz is quite similar in coverage to GSM 1800.

Higher frequencies are a disadvantage when it comes to coverage, but it is a decided advantage when it comes to capacity. Pico cells, covering e.g. one floor of a building, become possible, and the same frequency can be used for cells which are practically neighbours.

Cell service area may also vary due to interference from transmitting systems, both within and around that cell. This is true especially in CDMA based systems. The receiver requires a certain signal-to-noise ratio, and the transmitter should not send with too high transmission power in view to not cause interference with other transmitters. As the receiver moves away from the transmitter, the power received decreases, so the power control algorithm of the transmitter increases the power it transmits to restore the level of received power. As the interference (noise) rises above the received power from the transmitter, and the power of the transmitter cannot be increased any more, the signal becomes corrupted and eventually unusable. In CDMA-based systems, the effect of interference from other mobile transmitters in the same cell on coverage area is very marked and has a special name, *cell breathing*.

One can see examples of cell coverage by studying some of the coverage maps provided by real operators on their web sites or by looking at independently crowdsourced maps such as OpenSignal. In certain cases they may mark the site of the transmitter, in others it can be calculated by working out the point of strongest coverage.

A cellular repeater is used to extend cell coverage into larger areas. They range from wideband repeaters for consumer use in homes and offices to smart or digital repeaters for industrial needs.

Coverage Comparison of Different Frequencies

The following table shows the dependency of the coverage area of one cell on the frequency of a CDMA2000 network:

Frequency (MHz)	Cell radius (km)	Cell area (km²)	Relative Cell Count
450	48.9	7521	1
950	26.9	2269	3.3
1800	14.0	618	12.2
2100	12.0	449	16.2

Wireless Ad Hoc Network

An ad hoc network is a network that is composed of individual devices communicating with each other directly. The term implies spontaneous or impromptu construction because these networks often bypass the gatekeeping hardware or central access point such as a router. Many ad hoc networks are local area networks where computers or other devices are enabled to send data directly to one another rather than going through a centralized access point.

The idea of an ad hoc network is often unfamiliar to end users who have only seen small residential or business networks that use a typical router to send wireless signals to individual computers. However, the ad hoc network is being used quite a bit in new types of wireless engineering, although until recently it was a rather esoteric idea. For example, a mobile ad hoc network involves mobile devices communicating directly with one another. Another type of ad hoc network, the vehicular ad hoc network, involves placing communication devices in cars. Both of these are examples of ad hoc networks that use a large collection of individual devices to freely communicate without a kind of top-down or hierarchical communication structure.

Experts point out that for small local area networks, ad hoc networks can be cheaper to build because they don't require as much hardware. However, others make the point that a large number of devices can be difficult to manage without a larger and more concrete infrastructure. Tech leaders are looking at ways to enable more vibrant network functionality with these peer-to-peer networks.

Ad Hoc Wireless Network Features and uses

- Ad hoc networks are useful when you need to share files or other data directly with another computer but don't have access to a Wi-Fi network.

- More than one laptop can be connected to the ad hoc network, as long as all of the adapter cards are configured for ad-hoc mode and connect to the same SSID (service state identifier). The computers need to be within 100 meters of each other.

- If you are the person who sets up the ad hoc network, when you disconnect from the network, all the other users are also disconnected. An ad hoc network is deleted when everyone on it disconnects—which can be good or bad, depending on your view; it's truly a spontaneous network.

- You can use an ad hoc wireless network to share your computer's internet connection with another computer.

Early Work on MANET

In the early 1990s, Charles Perkins from SUN Microsystems USA, and Chai Keong Toh from Cambridge University separately started to work on a different Internet, that of a wireless ad hoc network. Perkins was working on the dynamic addressing issues. Toh worked on a new routing protocol, which was known as ABR – associativity-based routing. Perkins eventually proposed DSDV – Destination Sequence Distance Vector routing, which was based on distributed distance vector routing. Toh's proposal was an on-demand based routing, i.e. routes are discovered on-the-fly in real-time as and when needed. ABR was submitted to IETF as RFCs. ABR was implemented successfully into Linux OS on Lucent WaveLAN 802.11a enabled laptops and a practical ad hoc mobile network was therefore proven to be possible in 1999. Another routing protocol known as AODV was subsequently introduced and later proven and implemented in 2005. In 2007, David Johnson and Dave Maltz proposed DSR – Dynamic Source Routing.

Applications

The decentralized nature of wireless ad-hoc networks makes them suitable for a variety of applications where central nodes can't be relied on and may improve the scalability of networks compared to wireless managed networks, though theoretical and practical limits to the overall capacity of such networks have been identified. Minimal configuration and quick deployment make ad hoc networks suitable for emergency situations like natural disasters or military conflicts. The presence of dynamic and adaptive routing protocols enables ad hoc networks to be formed quickly. Wireless ad-hoc networks can be further classified by their applications:

Mobile Ad Hoc Networks (MANETs)

A mobile ad hoc network (MANET) is a continuously self-configuring, self-organizing, infrastructure-less network of mobile devices connected without wires. It is sometimes known as "on-the-fly" networks or "spontaneous networks".

Vehicular Ad Hoc Networks (VANETs)

VANETs are used for communication between vehicles and roadside equipment. Intelligent vehicular ad hoc networks (InVANETs) are a kind of artificial intelligence that helps vehicles to behave in intelligent manners during vehicle-to-vehicle collisions, accidents. Vehicles are using radio waves to communicate with each other, creating communication networks instantly on-the-fly while vehicles are moving on the roads.

Smart Phone Ad Hoc Networks (SPANs)

A SPAN leverages existing hardware (primarily Wi-Fi and Bluetooth) and software (protocols) in commercially available smartphones to create peer-to-peer networks without relying on cellular carrier networks, wireless access points, or traditional network infrastructure. Most recently, Apple's iPhone with version 7.0 iOS and higher have been enabled with multi-peer ad hoc mesh networking capability, in iPhones, allowing millions of smart phones to create ad hoc networks without relying on cellular communications. It has been claimed that this is going to "change the world".

Wireless Mesh Networks

Mesh networks take their name from the topology of the resultant network. In a fully connected mesh, each node is connected to every other node, forming a "mesh". A partial mesh, by contrast, has a topology in which some nodes are not connected to others, although this term is seldom in use. Wireless ad hoc networks can take the form of a mesh networks or others. A wireless ad hoc network does not have fixed topology, and its connectivity among nodes is totally dependent on the behavior of the devices, their mobility patterns, distance with each other, etc. Hence, wireless mesh networks are a particular type of wireless ad hoc networks, with special emphasis on the resultant network topology. While some wireless mesh networks (particularly those within a home) have relatively infrequent mobility and thus infrequent link breaks, other more mobile mesh networks require frequent routing adjustments to account for lost links. Google Home, Google Wi-Fi, and Google OnHub all support Wi-Fi mesh (i.e., Wi-Fi ad hoc) networking. Apple's AirPort allows the formation of wireless mesh networks at home, connecting various Wi-Fi devices together and providing good wireless coverage and connectivity at home.

Army Tactical MANETs

Army has been in need of "on-the-move" communications for a long time. Ad hoc mobile communications come in well to fulfill this need, especially its infrastructureless nature, fast deployment and operation. Military MANETs are used by military units with emphasis on rapid deployment, infrastructureless, all-wireless networks (no fixed radio towers), robustness (link breaks are no problem), security, range, and instant operation. MANETs can be used in army "hopping" mines, in platoons where soldiers communicate in foreign terrains, giving them superiority in the battlefield. Tactical MANETs can be formed automatically during the mission and the network "disappears" when the mission is over or commissioned. It is sometimes called "on-the-fly" wireless tactical network.

Air Force UAV Ad Hoc Networks

Unmanned aerial vehicle, is an aircraft with no pilot on board. UAVs can be remotely controlled (i.e., flown by a pilot at a ground control station) or can fly autonomously based on pre-programmed flight plans. Civilian usage of UAV include modeling 3D terrains, package delivery (Amazon), etc.

UAVs have also been used by US Air Force for data collection and situation sensing, without risking the pilot in a foreign unfriendly environment. With wireless ad hoc network technology embedded into the UAVs, multiple UAVs can communicate with each other and work as a team, collaboratively to complete a task and mission. If a UAV is destroyed by an enemy, its data can be quickly offloaded wirelessly to other neighboring UAVs. The UAV ad hoc communication network is also sometimes referred to UAV instant sky network.

Navy Ad Hoc Networks

Navy ships traditionally use satellite communications and other maritime radios to communicate with each other or with ground station back on land. However, such communications are restricted by delays and limited bandwidth. Wireless ad hoc networks enable ship-area-networks to be formed while at sea, enabling high speed wireless communications among ships, enhancing their sharing of imaging and multimedia data, and better co-ordination in battlefield operations. Some defense companies (such as Rockwell Collins and Rohde & Schwartz) have produced products that enhance ship-to-ship and ship-to-shore communications.

Wireless Sensor Networks

Sensors are useful devices that collect information related to a specific parameter, such as noise, temperature, humidity, pressure, etc. Sensors are increasingly connected via wireless to allow large scale collection of sensor data. With a large sample of sensor data, analytics processing can be used to make sense out of these data. The connectivity of wireless sensor networks rely on the principles behind wireless ad hoc networks, since sensors can now be deploy without any fixed radio towers, and they can now form networks on-the-fly. "Smart Dust" was one of the early projects done at U C Berkeley, where tiny radios were used to interconnect smart dust. More recently, mobile wireless sensor networks (MWSNs) have also become an area of academic interest.

Ad Hoc Home Smart Lighting

ZigBee is a low power form of wireless ad hoc networks that is now finding their way in home automation. Its low power consumption, robustness and extended range inherent in mesh networking can deliver several advantages for smart lighting in homes and in offices. The control includes adjusting dimmable lights, color lights, and color or scene. The networks allow a set or subset of lights to be controlled over a smart phone or via a computer. The home automation market is tipped to exceed $16 billion by 2019.

Ad Hoc Street Light Networks

Wireless ad hoc smart street light networks are beginning to evolve. The concept is to use wireless control of city street lights for better energy efficiency, as part of a smart city architectural feature. Multiple street lights form a wireless ad hoc network. A single gateway device can control up to 500 street lights. Using the gateway device, one can turn individual lights ON, OFF or dim them, as well as find out which individual light is faulty and in need of maintenance.

Ad Hoc Networked of Robots

Robots are mechanical systems that drive automation and perform chores that would seem difficult for man. Efforts have been made to co-ordinate and control a group of robots to undertake collaborative work to complete a task. Centralized control is often based on a "star" approach, where robots take turns to talk to the controller station. However, with wireless ad hoc networks, robots can form a communication network on-the-fly, i.e., robots can now "talk" to each other and collaborate in a distributed fashion. With a network of robots, the robots can communicate among themselves, share local information, and distributively decide how to resolve a task in the most effective and efficient way.

Disaster Rescue Ad Hoc Network

Another civilian use of wireless ad hoc network is public safety. At times of disasters (floods, storms, earthquakes, fires, etc.), a quick and instant wireless communication network is necessary. Especially at times of earthquakes when radio towers had collapsed or were destroyed, wireless ad hoc networks can be formed independently. Firemen and rescue workers can use ad hoc networks to communicate and rescue those injured. Commercial radios with such capability are available on the market.

Hospital Ad Hoc Network

Wireless ad hoc networks allow sensors, videos, instruments, and other devices to be deployed and interconnected wirelessly for clinic and hospital patient monitoring, doctor and nurses alert notification, and also making senses of such data quickly at fusion points, so that lives can be saved.

Challenges

Several books and works have revealed the technical and research challenges facing wireless ad hoc networks or MANETs. The advantages and disadvantages can be briefly summarized below:

Advantages

- Highly performing network.
- No expensive infrastructure must be installed
- Use of unlicensed frequency spectrum

- Quick distribution of information around sender

- No single point of failure.

Disadvantages

- All network entities may be mobile ⇒ very dynamic topology

- Network functions must have high degree of adaptability

- No central entities ⇒ operation in completely distributed manner.

Radios for Ad Hoc

Wireless ad hoc networks can operate over different types of radios. They can be UHF (300 – 3000 MHz), SHF (3 – 30 GHz), and EHF (30 – 300 GHz). Wi-Fi ad hoc uses the unlicensed ISM 2.4 GHz radios. They can also be used on 5.8 GHz radios.

Next generation Wi-Fi known as 802.11ax provides low delay, high capacity (up to 10Gbit/s) and low packet loss rate, offering 12 streams – 8 streams at 5 GHz and 4 streams at 2.4 GHz. IEEE 802.11ax uses 8x8 MU-MIMO, OFDMA, and 80 MHz channels. Hence, 802.11ax has the ability to form high capacity Wi-Fi ad hoc networks.

At 60 GHz, there is another form of Wi-Fi known as WiGi – wireless gigabit. This has the ability to offer up to 7Gbit/s throughput. Currently, WiGi is targeted to work with 5G cellular networks.

The higher the frequency, such as those of 300 GHz, absorption of the signal will be more predominant. Army tactical radios usually employ a variety of UHF and SHF radios, including those of VHF to provide a variety of communication modes. At the 800, 900, 1200, 1800 MHz range, cellular radios are predominant. Some cellular radios use ad hoc communications to extend cellular range to areas and devices not reachable by the cellular base station.

Protocol Stack

The challenges affecting MANETs span from various layers of the OSI protocol stack. The media access layer (MAC) has to be improved to resolve collisions and hidden terminal problems. The network layer routing protocol has to be improved to resolve dynamically changing network topologies and broken routes. The transport layer protocol has to be improved to handle lost or broken connections. The session layer protocol has to deal with discovery of servers and services.

A major limitation with mobile nodes is that they have high mobility, causing links to be frequently broken and reestablished. Moreover, the bandwidth of a wireless channel is also limited, and nodes operate on limited battery power, which will eventually be exhausted. Therefore, the design of a mobile ad hoc network is highly challenging, but this technology has high prospects to be able to manage communication protocols of the future.

The cross-layer design deviates from the traditional network design approach in which each layer of the stack would be made to operate independently. The modified transmission power will help that node to dynamically vary its propagation range at the physical layer. This is because the propagation distance is always directly proportional to transmission power. This information is passed from the physical layer to the network layer so that it can take optimal decisions in routing protocols. A major advantage of this protocol is that it allows access of information between physical layer and top layers (MAC and network layer).

Some elements of the software stack were developed to allow code updates *in situ*, i.e., with the nodes embedded in their physical environment and without needing to bring the nodes back into the lab facility. Such software updating relied on epidemic mode of dissemination of information and had to be done both efficiently (few network transmissions) and fast.

Routing

Routing in wireless ad hoc networks or MANETs generally falls into three categories, namely: (a) proactive routing, (b) reacting routing, and (c) hybrid routing.

Proactive Routing

This type of protocols maintains fresh lists of destinations and their routes by periodically distributing routing tables throughout the network. The main disadvantages of such algorithms are:

- Respective amount of data for maintenance.
- Slow reaction on restructuring and failures.

Example: Optimized Link State Routing Protocol (OLSR)

Distance Vector Routing

As in a fix net nodes maintain routing tables. Distance-vector protocols are based on calculating the direction and distance to any link in a network. "Direction" usually means the next hop address and the exit interface. "Distance" is a measure of the cost to reach a certain node. The least cost route between any two nodes is the route with minimum distance. Each node maintains a vector (table) of minimum distance to every node. The cost of reaching a destination is calculated using various route metrics. RIP uses the hop count of the destination whereas IGRP takes into account other information such as node delay and available bandwidth.

Reactive Routing

This type of protocol finds a route based on user and traffic demand by flooding the network with Route Request or Discovery packets. The main disadvantages of such algorithms are:

- High latency time in route finding.
- Excessive flooding can lead to network clogging.

However, clustering can be used to limit flooding. The latency incurred during route discovery is not significant compared to periodic route update exchanges by all nodes in the network.

Example: Ad hoc On-Demand Distance Vector Routing (AODV)

Flooding

Is a simple routing algorithm in which every incoming packet is sent through every outgoing link except the one it arrived on. Flooding is used in bridging and in systems such as Usenet and peer-to-peer file sharing and as part of some routing protocols, including OSPF, DVMRP, and those used in wireless ad hoc networks.

Hybrid Routing

This type of protocol combines the advantages of *proactive* and *reactive routing*. The routing is initially established with some proactively prospected routes and then serves the demand from additionally activated nodes through reactive flooding. The choice of one or the other method requires predetermination for typical cases. The main disadvantages of such algorithms are:

- Advantage depends on number of other nodes activated.
- Reaction to traffic demand depends on gradient of traffic volume.

Example: Zone Routing Protocol (ZRP)

Position-based Routing

Position-based routing methods use information on the exact locations of the nodes. This information is obtained for example via a GPS receiver. Based on the exact location the best path between source and destination nodes can be determined.

Example: "Location-Aided Routing in mobile ad hoc networks" (LAR)

Technical Requirements for Implementation

An ad hoc network is made up of multiple "nodes" connected by "links."

Links are influenced by the node's resources (e.g., transmitter power, computing power and memory) and behavioral properties (e.g., reliability), as well as link properties (e.g. length-of-link and signal loss, interference and noise). Since links can be connected or disconnected at any time, a functioning network must be able to cope with this dynamic restructuring, preferably in a way that is timely, efficient, reliable, robust, and scalable.

The network must allow any two nodes to communicate by relaying the information via other nodes. A "path" is a series of links that connects two nodes. Various routing methods use one or two paths between any two nodes; flooding methods use all or most of the available paths.

Medium-access Control

In most wireless ad hoc networks, the nodes compete for access to shared wireless medium, often resulting in collisions (interference). Collisions can be handled using centralized scheduling or distributed contention access protocols. Using cooperative wireless communications improves immunity to interference by having the destination node combine self-interference and other-node interference to improve decoding of the desired signals.

Software Reprogramming

Large-scale ad hoc wireless networks may be deployed for long periods of time. During this time the requirements from the network or the environment in which the nodes are deployed may change. This can require modifying the application executing on the sensor nodes, or providing the application with a different set of parameters. It may be very difficult to manually reprogram the nodes because of the scale (possibly hundreds of nodes) and the embedded nature of the deployment, since the nodes may be located in places that are difficult to access physically. Therefore, the most relevant form of reprogramming is *remote multihop reprogramming* using the wireless medium which reprograms the nodes as they are embedded in their sensing environment. Specialized protocols have been developed for the embedded nodes which minimize the energy consumption of the process as well as reaching the entire network with high probability in as short a time as possible.

Mathematical Models

The traditional model is the random geometric graph. Early work included simulating ad hoc mobile networks on sparse and densely connected topologies. Nodes are firstly scattered in a constrained physical space randomly. Each node then has a predefined fixed cell size (radio range). A node is said to be connected to another node if this neighbor is within its radio range. Nodes are then moved (migrated away) based on a random model, using random walk or brownian motion. Different mobility and number of nodes present yield different route length and hence different number of multi-hops.

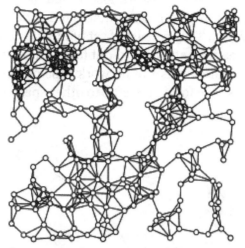

A randomly constructed geometric graph drawn inside a square

These are graphs consisting of a set of nodes placed according to a point process in some usually bounded subset of the n-dimensional plane, mutually coupled according to a boolean probability mass function of their spatial separation. The connections between nodes may have different weights to model the difference in channel attenuations. One can then study network observables (such as connectivity, centrality or the degree distribution) from a graph-theoretic perspective. One can further study network protocols and algorithms to improve network throughput and fairness.

Security

Most wireless ad hoc networks do not implement any network access control, leaving these networks vulnerable to resource consumption attacks where a malicious node injects packets into the network with the goal of depleting the resources of the nodes relaying the packets.

To thwart or prevent such attacks, it was necessary to employ authentication mechanisms that ensure that only authorized nodes can inject traffic into the network. Even with authentication, these networks are vulnerable to packet dropping or delaying attacks, whereby an intermediate node drops the packet or delays it, rather than promptly sending it to the next hop.

Simulation

One key problem in wireless ad hoc networks is foreseeing the variety of possible situations that can occur. As a result, modeling and simulation (M&S) using extensive parameter sweeping and what-if analysis becomes an extremely important paradigm for use in ad hoc networks. Traditional M&S tools include OPNET, and NetSim.

Mobile Ad Hoc Network

MANET stands for Mobile adhoc Network also called as wireless adhoc network or adhoc wireless network that usually has a routable networking environment on top of a Link Layer ad hoc network.. They consist of set of mobile nodes connected wirelessly in a self configured, self healing network without having a fixed infrastructure. MANET nodes are free to move randomly as the network topology changes frequently. Each node behave as a router as they forward traffic to other specified node in the network.

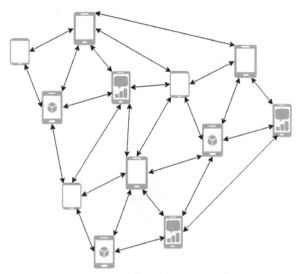

Figure- Mobile Ad hoc network

MANET may operate as standalone fashion or they can be the part of larger internet. They form highly dynamic autonomous topology with the presence of one or multiple different transceivers between nodes. The main challenge for the MANET is to equipped each devices to continuously maintain the information required to properly route traffic. MANETs consist of a peer-to-peer, self-forming, self-healing network MANET's circa 2000-2015 typically communicate at radio frequencies (30MHz-5GHz). This can be used in road safety, ranging from sensors for environment, home, health, disaster rescue operations, air/land/navy defense, weapons, robots, etc.

Characteristics of MANET

- Dynamic Topologies: Network topology which is typically multihops, may change randomly and rapidly with time, it can form unidirectional or bi-directional links.

- Bandwidth constrained, variable capacity links: Wireless links usually have lower reliability, efficiency, stability and capacity as compared to wired network. The throughput of wireless communication is even less than a radio's maximum transmission rate after dealing with the constraints like multiple access, noise, interference conditions, etc.

- Autonomous Behavior: Each nodes can act as a host and router, which shows its autonomous behavior.

- Energy Constrained Operation: As some or all the nodes rely on batteries or other exhaustible means for their energy.Mobile nodes are characterized with less memory, power and light weight features.

- Limited Security: Wireless network are more prone to security threats. A centralized firewall is absent due to its distributed nature of operation for security, routing and host configuration.

- Less Human Intervention: They require minimum human intervention to configure the network, therefore they are dynamically autonomous in nature.

Types

- Vehicular ad hoc networks (VANETs) are used for communication between vehicles and roadside equipment. Intelligent vehicular ad hoc networks (InVANETs) are a kind of artificial intelligence that helps vehicles to behave in intelligent manners during vehicle-to-vehicle collisions, accidents.

- Smart phone ad hoc networks (SPANs) leverage the existing hardware (primarily Bluetooth and Wi-Fi) in commercially available smart phones to create peer-to-peer networks without relying on cellular carrier networks, wireless access points, or traditional network infrastructure. SPANs differ from traditional hub and spoke networks, such as Wi-Fi Direct, in that they support multi-hop relays and there is no notion of a group leader so peers can join and leave at will without destroying the network.

- Internet-based mobile ad-hoc networks (iMANETs) is a type of wireless ad hoc network that supports Internet protocols such as TCP/UDP and IP. The network uses a network-layer routing protocol to link mobile nodes and establish routes distributedly and automatically.

- Hub-Spoke MANET – Multiple sub-MANETs may be connected in a classic Hub-Spoke VPN to create a geographically distributed MANET. In such type of networks normal ad hoc routing algorithms does not apply directly. One implementation of this is Persistent System's CloudRelay.

- Military or tactical MANETs are used by military units with emphasis on data rate, real-time requirement, fast re-routing during mobility, data security, radio range, and integration with existing systems. Common radio waveforms include the US Army's JTRS SRW and Persistent System's WaveRelay.

- Flying ad hoc networks (FANETs) are composed of unmanned aerial vehicles, allowing great mobility and providing connectivity to remote areas.

Advantages and Disadvantages in Wireless Communication Networks

The obvious appeal of MANETs is that the network is decentralised and nodes/devices are mobile, that is to say there is no fixed infrastructure which provide the possibility for numerous applications in different areas such as environmental monitoring disaster relief – and military communications. Since the early 2000s interest in MANETs has greatly increased which, in part, is due to the fact mobility can improve network capacity, shown by Grossglauser and Tse along with the introduction of new technologies.

One main advantage to a decentralised network is that they are typically more robust than centralised networks due to the multi-hop fashion in which information is relayed. For example, in the cellular network setting, a drop in coverage occurs if a base station stops working, however the chance of a single point of failure in a MANET is reduced significantly since the data can take multiple paths. Since the MANET architecture evolves with time it has the potential to resolve issues such as isolation/disconnection from the network. Further advantages of MANETS over networks with a fixed topology include flexibility (an ad hoc network can be created anywhere with mobile devices), scalability (you can easily add more nodes to the network) and lower administration costs (no need to build an infrastructure first).

With these positives follow some obvious draw backs in network performance. With a time evolving network it is clear we should expect variations in network performance due to no fixed architecture (no fixed connections). Furthermore, since network topology determines interference and thus connectivity, the mobility pattern of devices within the network will impact on network performance, possibly resulting in data having to be resent a lot of times (increased delay) and finally allocation of network resources such as power remains unclear. Finally, finding a model that accurately represents human mobility whilst remaining mathematically tractable remains an open problem due to the large range of factors that influence it. Some typical models used include the random walk, random waypoint and levy flight models.

Applications

Mobile ad hoc networks can be used in many applications, ranging from sensors for environment, vehicular ad hoc communications, road safety, health, home, peer-to-peer messaging, disaster rescue operations, air/land/navy defense, weapons, robots, etc.

Simulations

There are several ways to study MANETs. One solution is the use of simulation tools like OPNET, NetSim, NS2, OMNeT++ and NS3. A comparative study of various simulators for VANETs reveal that factors such as constrained road topology, multi-path fading and roadside obstacles, traffic flow models, trip models, varying vehicular speed and mobility, traffic lights, traffic congestion, drivers' behavior, etc., have to be taken into consideration in the simulation process to reflect realistic conditions.

Data Monitoring and Mining

MANETS can be used for facilitating the collection of sensor data for data mining for a variety of applications such as air pollution monitoring and different types of architectures can be used for such applications. It should be noted that a key characteristic of such applications is that nearby sensor nodes monitoring an environmental feature typically register similar values. This kind of data redundancy due to the spatial correlation between sensor observations inspires the techniques for in-network data aggregation and mining. By measuring the spatial correlation between data sampled by different sensors, a wide class of specialized algorithms can be developed to develop more efficient spatial data mining algorithms as well as more efficient routing strategies. Also, researchers have developed performance models for MANET to apply queueing theory.

Smart Phone Ad Hoc Network

While the cellphone network in Haiti survived the devastating earthquake in 2010, the added load of international aid workers who arrived in the aftermath caused it to crash. Josh Thomas and Jeff Robble, both working at Mitre, saw this problem and created a working prototype backup network using only the Wi-Fi chips on Android smartphones. This capability won't be shipped on new mobile phones anytime soon, but it is a really interesting open innovation project to understand and follow, and for some an Android project to which they might contribute.

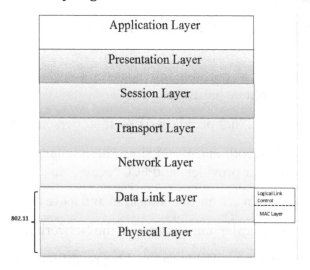

The Smart Phone Ad-Hoc Networks (SPAN) project reconfigures the onboard Wi-Fi chip of a smartphone to act as a Wi-Fi router with other nearby similarly configured smartphones, creating an ad-hoc mesh network. These smartphones can then communicate with one another without an operational carrier network. SPAN intercepts all communications at the Global Handset Proxy. So applications such as VoIP, Twitter, email etc., work normally.

The source from the Linux Wireless Extension API was merged into the Android kernel source and compiled. The modified version of Android was used to root specific models of Android smartphones to expose and harness the ad-hoc routing features of the on-board Wi-Fi chip to enable this intercept.

It is really a framework for further research to refine how to build the special case of an ad-hoc mesh network. SPAN's routing module is designed to be plug-and-play so it can be easily replaced. Researchers and developers interested in experimenting with new routing protocols save months of man-hours needed to build the entire app by using the SPAN framework.

The current version can be toggled between widely adopted routing protocols OLSRd and Dijkstra to test the differences in the performance of network discovery and routing. In testing SPAN, the limits of these routing protocols were discovered. Network discovery floods a network with "hello" packets so a routing table can be built. This type of discovery works well in static networks because the amount of bandwidth used for discovery is limited to infrequent changes in the network.

Features

- Capable of going off-grid and enabling peer-to-peer communications without relying on cellular carrier networks, wireless access points, or traditional network infrastructure.

- Optional Internet access through gateway devices, such as mobile hotspots in the mesh.

- Optional stationary or portable infrastructures such as routers, mesh extenders, or other non-phone hardware.

- Leverage the devices that people carry on their person and use every day.

- Primarily use Bluetooth and Wi-Fi since the cellular spectrum is licensed and controlled by cellular providers and FCC regulations.

- Set up and tear down are on-demand. Join and leave at will.

- Routing protocol may be implemented at the Network Layer or Link Layer.

- Often requires rooting a device and making modifications to the operating system, kernel, or drivers.

Threats to Telcos

The ad hoc networking technology operating on Wi-Fi ad hoc mode, at the unlicensed ISM band of 2.4 GHz may result in profit loss by cellular carriers since ISM band is free and unlicensed while cellular carriers operate on licensed band at 900 MHz, 1200 MHz, 1800 MHz, etc. This has the potential to threaten telecommunication operators (telcos). Smart phone mobile ad hoc networks can operate independently and allow communications among smart phones users without the need for any 3G or 4G LTE signals to be present. Wi-Fi ad hoc mode was first implemented on Lucent WaveLan 802.11a/b on laptop computers. The technology success is now carried forward and used in smart phones, since Wi-Fi is present and embedded in all smart phones today.

Important Applications

- Developing nations where network infrastructure doesn't exist.

- Protests where government entities monitor or disable network infrastructure.

- Natural disasters or terrorist incidents where existing network infrastructure is overloaded, destroyed, or compromised.

- Temporary Large-scale events such as festivals where huge scale is needed for short period of time

Below are some of the recent applications of smart phone ad hoc networks in real life:

- 2014 – in Iraq following government restrictions on internet use, users use the technology to communicate

- 2014 – Hong Kong Protests in China used Firechat to communicate

- 2015 – Leaders of Anti-government protests in Russia in December 2014 urged their followers to install firechat

Apple Multipeer Connectivity

In Apple Inc. iPhones released with iOS version 7.0 and higher, multipeer connectivity APIs (application programmable interfaces) are enabled and provided to allow Apple iPhones to operate in peer-to-peer ad hoc mesh networking mode. This means iPhones can now talk to each other without using a cellular signal or connection. Currently, Apple uses multipeer to allow one to send photos and large files (up to GB) to peers. This application is called AirDrop and has been gaining popularity. With 700+ millions of iPhones being used globally, ad hoc peer-to-peer networks will gain pervasive presence almost instantly.

Millions of teens and school kids in the United States have benefited and been using smart phone ad hoc networks to chat and communicate, without the need for Wi-Fi or cellular signals. The penetration will continue to grow, expanding to teens outside the US.

Android Phone Ad Hoc Networks

By merging Linux Wireless Extension API with the Android kernel source and recompiling, the modified version of Android can enable the ability to harness the ad hoc routing, communications and networking features of the onboard Wi-Fi chip. This empowers millions of Android phones (example Google Pixel phones) with ad hoc mesh connectivity and networking capabilities.

Projects

- The Serval Project
- Commotion Wireless
- FireChat

Device Manufacturer Support

- iOS Multipeer Connectivity Framework

Vehicular Ad Hoc Network

Vehicular ad hoc networks (VANETs) have been quite a hot research area in the last few years. Due to their unique characteristics such as high dynamic topology and predictable mobility, VANETs attract so much attention of both academia and industry. In this paper, we provide an overview of the main aspects of VANETs from a research perspective. This paper starts with the basic architecture of networks, then discusses three popular research issues and general research methods, and ends up with the analysis on challenges and future trends of VANETs.

Recently, with the development of vehicle industry and wireless communication technology, vehicular ad hoc networks are becoming one of the most promising research fields.

VANETs which use vehicles as mobile nodes are a subclass of mobile ad hoc networks (MANETs) to provide communications among nearby vehicles and between vehicles and nearby roadside equipment but apparently differ from other networks by their own characteristics. Specifically, the nodes (vehicles) in VANETs are limited to road topology while moving, so if the road information is available, we are able to predict the future position of a vehicle; what is more, vehicles can afford significant computing, communication, and sensing capabilities as well as providing continuous transmission power themselves to support these functions .

However, VANETs also come with several challenging characteristics, such as potentially large scale and high mobility. Nodes in the vehicular environment are much more dynamic because most cars usually are at a very high speed and change their position constantly. The high mobility also leads to a dynamic network topology, while the links between nodes connect and disconnect very often. Besides, VANETs have a potentially large scale which can include many participants and extend over the entire road network .

Special Characteristics of VANET

Though VANET could be treated as a subgroup of Mobile Ad Hoc Networks (MANETs) and it is still necessary to consider VANETs as a distinct research field, especially in the light of security provisioning. The unique characteristics of VANET include features:

High Dynamic Topology: The speed and choice of path defines the dynamic topology of VANET. If we assume two vehicles moving away from each other with a speed of 60 km/h (25 m/s) and if the transmission range is about 250m, then the link between these two vehicles will last for only 5 seconds (250m). This defines its highly dynamic topology.

Frequent Disconnected Network: The above feature necessitates that in about every 5 seconds or so, the nodes needed another link with nearby vehicle to maintain seamless connectivity. But in case of such failure, particularly in case of low vehicle density zone, frequent disruption of network connectivity will occur. Such problems are at times addressed by road-side deployment of relay nodes.

Mobility Modeling and Prediction : The above features for connectivity therefore needed the knowledge of node positions and their movements which is as such very difficult to predict keeping in view the nature and pattern of movement of each vehicle. Nonetheless, a mobility model and node prediction based on study of predefined roadways model and vehicle speed is of paramount importance for effective network design.

Communication Environment:

The mobility model highly varies from highways to that of city environment. The node prediction design and routing algorithm also therefore need to adapt for these changes. Highway mobility model is essentially a one-dimensional model which is rather simple and easy to predict. But for city mobility model, street structure, variable node density, presence of buildings and trees behave as obstacles to even small distance communication make the model application very complex and difficult.

Unlimited Transmission Power:

The node (vehicle) itself can provide continuous power to computing and communication devices.

Hard Delay Constraints: The safety aspect (such as accidents, brake event) of VANET application warrants on time delivery of message to relevant nodes. It simply cannot compromise with any hard data delay in this regard. Therefore, high data rates are not as important as an issue for VANET as overcoming the issues of hard delay constraints.

Interaction with onboard Sensors: These sensors help in providing node location and their movement nature that are used for effective communication link and routing purposes.

Higher Computational Capability: Indeed, operating vehicles can afford significant computing, communication and sensing capabilities.

Rapidly Changing Network Topology: Due to high node mobility, the network topology in VANET tends to change frequently.

Potentially Unbounded Network Size: VANETs could involve the vehicles in one city, several cities or even a country. Thus, it is necessary to make any protocols for VANET is scalable in order to be practical.

Anonymous Addressee: Most applications in VANETs require identification of the vehicles in a certain region, instead of the specific vehicles. This may help protect node privacy in VANETs.

Time-Sensitive Data Exchange: Most safety related applications require data packet transmission in a timely manner. Thus, any security schemes cannot harm the network performance of VANETs.

Potential Support from Infrastructure: Unlike common MANETs, VANETs can actually take advantage of infrastructure in the future. This property has to be considered to make any protocols and a scheme for VANET is better.

Abundant Resources: VANET nodes have abundant energy and computation resources. This allows schemes involving usage of resource demanding techniques such as ECDSA, RSA etc.

Better Physical Protection: VANET nodes are better protected than MANETs. Thus, Nodes are more difficult to compromise which is also good news for security provisioning in VANETs.

Partitioned Network:

Vehicular networks will be frequently partitioned. The dynamic nature of traffic may result in large inter vehicle gaps in sparsely populated scenarios and hence in several isolated clusters of nodes.

Technology

VANETs can use any wireless networking technology as their basis. The most prominent are short range radio technologies like WLAN (either standard Wi-Fi or ZigBee). In addition, cellular technologies or LTE can be used for VANETs.The latest technology for this wireless networking is visible light communication[VLC](Infrared transmission and reception).

Simulations

Prior to the implementation of VANETs on the roads, realistic simulations of VANETs using a combination of Urban Mobility simulation and Network simulation is are necessary. Typically open source simulator like SUMO (which handles road traffic simulation) is combined with a network simulator like NetSim (TETCOS), to study the performance of VANETs. A study of cooperative automated driving using Webots and NS3 simulators before real road test has been performed in the context of the Autonet2030 European project.

Standards

Major standardization of VANET protocol stacks is taking place in the U.S., in Europe, and in Japan, corresponding to their dominance in the automotive industry.

In the U.S., the IEEE 1609 WAVE Wireless Access in Vehicular Environments protocol stack builds on IEEE 802.11p WLAN operating on seven reserved channels in the 5.9 GHz frequency band. The WAVE protocol stack is designed to provide multi-channel operation (even for vehicles equipped with only a single radio), security, and lightweight application layer protocols. Within the IEEE Communications Society, there is a Technical Subcommittee on Vehicular Networks & Telematics Applications (VNTA). The charter of this committee is to actively promote technical activities in the field of vehicular networks, V2V, V2R and V2I communications, standards, communications-enabled road and vehicle safety, real-time traffic monitoring, intersection management technologies, future telematics applications, and ITS-based services.

Applications of VANET

Vehicular ad-hoc network applications range from road safety applications oriented to the vehicle or to the driver, to entertainment and commercial applications for passengers, making use of a plethora of cooperating technologies. Thus, we have divided the applications into two major categories:

Safety-Related Applications: These applications contain safety related applications such as collision avoidance and cooperative driving (e.g., for lane merging). The common characteristic of this category is the relevance to life-critical situations where the existence of a service may prevent life-endangering accidents. Hence, the security of this category is mandatory, since the proper operation of any of these applications should be guaranteed even in the presence of attackers.

Other Applications: It includes traffic optimization, payment services (e.g. toll collection), location-based services (e.g. finding the closest fuel station) and entertainment information (e.g. Internet access). Obviously, security is also required in this application category especially in the case of payment services. But in this paper we focus on the security aspects of safety-related applications because they are the most specific to the automotive domain because they raise the most challenging problems. VANET would support life-critical safety applications, safety warning applications, electronic toll collections, Internet access, group communications, roadside service finder etc.

Life-Critical Safety Applications: It provides Intersection Collision Warning/Avoidance, Cooperative Collision Warning etc. In the MAC Layer, the Life-Critical Safety Applications can access the DSRC (Dedicated Short-Range Communications) control channel and other channels with the highest priority. The messages can be broadcasted to all the nearby VANET nodes.

Safety Warning Applications: Safety Warning Applications contain Work Zone Warning, Transit Vehicle Signal Priority etc. The differences between Life-Critical Safety Applications and Safety Warning Applications are the allowable latency requirements while the Life-Critical Safety Applications usually require the messages to be delivered to the nearby nodes within 100 milliseconds, the Safety Warning Applications can afford up to 1000 milliseconds. In the MAC Layer, the Safety Warning Applications can access the DSRC (Dedicated Short-Range Communications) control channel and the other channels with the 2nd highest priority. The messages can be broadcasted to all the nearby VANET nodes.

Electronic Toll Collections (ETCs):

Each vehicle can pay the toll electronically when it passes through a Toll Collection Point (a special RSU) without stopping. The Toll Collection Point will scan the Electrical License Plate at the OBU (on-board unit) of the vehicle and issue a receipt message to the vehicle including the amount of the toll, the time and the location of the Toll Collection Point. In the MAC layer, the Electronic Toll Collections priority. It should be a direct one-hop wireless link between the Toll Collection Point and the vehicle.

Internet Access: Future vehicles will be equipped with the capability so that the passages on the vehicles can connect to the Internet. In the MAC layer, the Internet Access applications can use DSRC service channels except the control channel, with the lowest priority comparing with the previous applications. In the network layer, it is used to support VANET Internet access, a straight forward method is to provide a unicast connection between the OBU of the vehicle and a RSU, which has the link towards the Internet.

Group Communications: Many drivers may share some common interests when they are on the same road to the same direction, so they can use the VANET Group Communication functions. In the MAC layer, the Group Communications can use DSRC service channels except the control channel with the lowest priority comparing with the safety related applications and ETCs. In the network layer, It is used to support such application scenario in which multicast is the key technology. In the past, Internet multicast has not been successful due to its complexity because Internet multicast requires global deployment, which is virtually impossible. In a VANET, however, since all nodes are located in a relatively local area, implementing such group communication becomes possible. There are many Roadside Service Finders. For example, find restaurants, gas stations etc., in the nearby area along the road. A Roadside Services Database will be installed in the local area that connected to the corresponding RSUs (Road Side Units). In the MAC layer, the Roadside Services Finder application can use DSRC service channels except the control channel, with the lowest priority comparing with the safety related applications and ETCs. Each vehicle can issue a Service Finder Request message that can be routed to the nearest RSU and a Service Finder Response message that can be routed back to the vehicle. In short, the application layer requirements must be addressed in the MAC layer and network layer design. In the next section, we provide a network design framework to satisfy the above applications while providing sufficient security.

DAHNI (Driver Ad Hoc Networking Infrastructure): DAHNI provides numerous possibilities to revolutionize the automotive and transportation industry of the future. For example, data captured by DAHNI, when properly aggregated, can be fed into the traffic monitoring and flow control system for real-time traffic management. Alternatively, such information can be archived for off-line analysis to understand traffic bottlenecks and devise techniques to alleviate traffic congestion.

Dedicated Short-Range Communications (DSRC) :

DSRC system has emerged in North America, where 75 MHz of spectrum was approved by the U.S. FCC (Federal Communication Commission) in 2003 for such type of communication that mainly targets vehicular networks. On the other hand, the Car-to-Car Communication Consortium (C2C-CC) has been initiated in Europe by car manufacturers and automotive OEMs (original equipment manufacturers), with the main objective of increasing road traffic safety and efficiency by means of inter vehicle communication. IEEE is also advancing within the IEEE 1609 family of standards for wireless access in vehicular environments (WAVE).

Future of VANET

It's basically a network of cars that are in constant communication. Each car knows where it is, where it's going and basically any other quantity that it can measure.

Not only has every car been made "self aware", it can also communicate with any other car on the road. Just take a minute to think of all the possibilities. The first one that comes to mind is road safety. If all cars know where all other cars are, cars headed for an imminent crash can warn their respective drivers and even apply autonomous control to avoid accidents. Not only drivers can be informed about delays to their destination, but traffic lights can be connected to the communication network to most efficiently route traffic in real time. This creates a system where the road network always serves the current needs of the users on the network.

VANET is a promising wireless communication technology for improving highway safety and information services. In this paper, we propose a secure and application oriented network design framework in which the requirements of potential VANET applications are taken into account. We also study several applications as a promising tool for monitoring the physical world with wireless sensor that can sense, process and communicate. VANET applications range from safety and crash avoidance to Internet access and multimedia. We believe that our study can provide a guideline for more practical VANET.

References

- C. Siva Ram Murthy and B. S. Manoj, Ad hoc Wireless Networks: Architectures and Protocols, Prentice Hall PTR, May 2004. ISBN 978-0-13-300706-0

- Chai Keong Toh (2002). "Ad Hoc Mobile Wireless Networks: Protocols and Systems 1st Edition". Prentice Hall PTR. Retrieved 2016-04-20

- J. Burchfiel; R. Tomlinson; M. Beeler (May 1975). Functions and structure of a packet radio station (PDF). National Computer Conference and Exhibition. pp. 245–251. doi:10.1145/1499949.1499989

- "History of Wireless". Johns Hopkins Bloomberg School of Public Health. Archived from the original on 2007-02-10. Retrieved 2007-02-17

- Martinez; Toh; Cano; et al. (2009). "A survey and comparative study of simulators for vehicular ad hoc networks (VANETs)". Wireless Communications Journal. 11 (7): 813–828. doi:10.1002/wcm.859

- Guowang Miao; Jens Zander; Ki Won Sung; Ben Slimane (2016). Fundamentals of Mobile Data Networks. Cambridge University Press. ISBN 1107143217

- Morteza M. Zanjireh; Hadi Larijani (May 2015). A Survey on Centralised and Distributed Clustering Routing Algorithms for WSNs (PDF). IEEE 81st Vehicular Technology Conference. Glasgow, Scotland. doi:10.1109/VTCSpring.2015.7145650

- Ma, Y.; Guo, Y.; Tian, X.; Ghanem, M. (2011). "Distributed Clustering-Based Aggregation Algorithm for Spatial Correlated Sensor Networks". IEEE Sensors Journal. 11 (3): 641. Bibcode:2011ISenJ..11..641M. doi:10.1109/JSEN.2010.2056916

- "Obstacle Management in VANET using Game Theory and Fuzzy Logic Control". International Journal on Communication. 4 (1). June 2013. Retrieved 30 August 2013

3

Wireless Sensor Network

Wireless sensor networks are a group of sensors that are spatially dispersed and meant for recording and monitoring crucial information of the physical environment, such as temperature, pollution levels, wind, humidity, etc. The aim of this chapter is to explore the central aspects of wireless sensor networks, such as sensor grid and node, location estimation, visual and virtual sensor network, among others.

A wireless sensor network (WSN) is a wireless network consisting of spatially distributed autonomous devices using sensors to monitor physical or environmental conditions. A WSN system incorporates a gateway that provides wireless connectivity back to the wired world and distributed nodes. The wireless protocol you select depends on your application requirements. Some of the available standards include 2.4 GHz radios based on either IEEE 802.15.4 or IEEE 802.11 (Wi-Fi) standards or proprietary radios, which are usually 900 MHz.

A Wireless sensor network can be defined as a network of devices that can communicate the information gathered from a monitored field through wireless links. The data is forwarded through multiple nodes, and with a gateway, the data is connected to other networks like wireless Ethernet

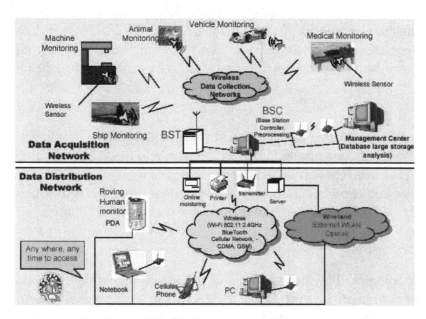

Wireless Sensor Networks

WSN is a wireless network that consists of base stations and numbers of nodes (wireless sensors).These networks are used to monitor physical or environmental conditions like sound, pressure, temperature and co-operatively pass data through the network to a main location as shown in the figure.

WSN Network Topologies

For radio communication networks, the structure of a WSN includes various topologies like the ones given below.

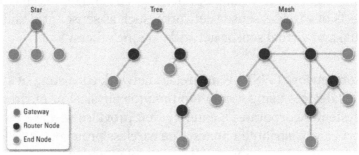

Wireless Sensor Network Topologies

Star Topologies

Star topology is a communication topology, where each node connects directly to a gateway. A single gateway can send or receive a message to a number of remote nodes. In star topologies, the nodes are not permitted to send messages to each other. This allows low-latency communications between the remote node and the gateway (base station).

Due to its dependency on a single node to manage the network, the gateway must be within the radio transmission range of all the individual nodes. The advantage includes the ability to keep the remote nodes' power consumption to a minimum and simply under control. The size of the network depends on the number of connections made to the hub.

Tree Topologies

Tree topology is also called as cascaded star topology. In tree topologies, each node connects to a node that is placed higher in the tree, and then to the gateway. The main advantage of the tree topology is that the expansion of a network can be easily possible, and also error detection becomes easy. The disadvantage with this network is that it relies heavily on the bus cable; if it breaks, all the network will collapse.

Mesh Topologies

The Mesh topologies allow transmission of data from one node to another, which is within its radio transmission range. If a node wants to send a message to another node, which is out of radio communication range, it needs an intermediate node to forward the message to the desired node. The advantage with this mesh topology includes easy isolation and detection of faults in the network. The disadvantage is that the network is large and requires huge investment.

Types of WSNs (Wireless Sensor Networks)

Depending on the environment, the types of networks are decided so that those can be deployed underwater, underground, on land, and so on. Different types of WSNs include:

1. Terrestrial WSNs
2. Underground WSNs
3. Underwater WSNs
4. Multimedia WSNs
5. Mobile WSNs

1. Terrestrial WSNs

Terrestrial WSNs are capable of communicating base stations efficiently, and consist of hundreds to thousands of wireless sensor nodes deployed either in unstructured (ad hoc) or structured (Preplanned) manner. In an unstructured mode, the sensor nodes are randomly distributed within the target area that is dropped from a fixed plane. The preplanned or structured mode considers optimal placement, grid placement, and 2D, 3D placement models.

In this WSN, the battery power is limited; however, the battery is equipped with solar cells as a secondary power source. The Energy conservation of these WSNs is achieved by using low duty cycle operations, minimizing delays, and optimal routing, and so on.

2. Underground WSNs

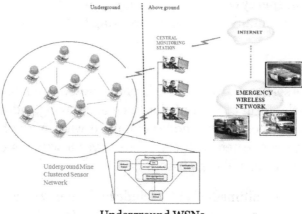

Underground WSNs

The underground wireless sensor networks are more expensive than the terrestrial WSNs in terms of deployment, maintenance, and equipment cost considerations and careful planning. The WSNs networks consist of a number of sensor nodes that are hidden in the ground to monitor underground conditions. To relay information from the sensor nodes to the base station, additional sink nodes are located above the ground.

The underground wireless sensor networks deployed into the ground are difficult to recharge. The sensor battery nodes equipped with a limited battery power are difficult to recharge. In addition to this, the underground environment makes wireless communication a challenge due to high level of attenuation and signal loss.

3. Under Water WSNs

More than 70% of the earth is occupied with water. These networks consist of a number of sensor nodes and vehicles deployed under water. Autonomous underwater vehicles are used for gathering data from these sensor nodes. A challenge of underwater communication is a long propagation delay, and bandwidth and sensor failures.

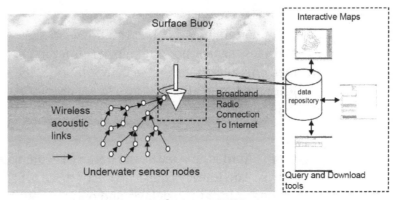

Under Water WSNs

Under water WSNs are equipped with a limited battery that cannot be recharged or replaced. The issue of energy conservation for under water WSNs involves the development of underwater communication and networking techniques.

4. Multimedia WSNs

Muttimedia wireless sensor networks have been proposed to enable tracking and monitoring of events in the form of multimedia, such as imaging, video, and audio. These networks consist of low-cost sensor nodes equipped with micrpphones and cameras. These nodes are interconnected with each other over a wireless connection for data compression, data retrieval and correlation.

The challenges with the multimedia WSN include high energy consumption, high bandwidth requirements, data processing and compressing techniques. In addition to this, multimedia contents require high bandwidth for the contents to be delivered properly and easily.

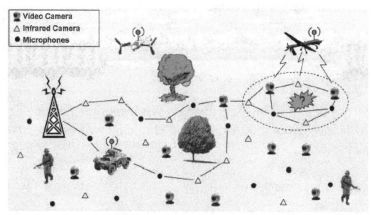
Multimedia WSNs

5. Mobile WSNs

These networks consist of a collection of sensor nodes that can be moved on their own and can be interacted with the physical environment. The mobile nodes have the ability to compute sense and communicate.

The mobile wireless sensor networks are much more versatile than the static sensor networks. The advantages of MWSN over the static wireless sensor networks include better and improved coverage, better energy efficiency, superior channel capacity, and so on.

Application

Area Monitoring

Area monitoring is a common application of WSNs. In area monitoring, the WSN is deployed over a region where some phenomenon is to be monitored. A military example is the use of sensors detect enemy intrusion; a civilian example is the geo-fencing of gas or oil pipelines.

Health Care Monitoring

The sensor networks for medical applications can be of several types: implanted, wearable, and environment-embedded. The implantable medical devices are those that are inserted inside human body. Wearable devices are used on the body surface of a human or just at close proximity of the user. Environment-embedded systems employ sensors contained in the environment. Possible applications include body position measurement, location of persons, overall monitoring of ill patients in hospitals and at homes. Devices embedded in the environment track the physical state of a person for continuous health diagnosis, using as input the data from a network of depth cameras, a sensing floor, or other similar devices. Body-area networks can collect information about an individual's health, fitness, and energy expenditure. In health care applications the privacy and authenticity of user data has prime importance. Especially due to the integration of sensor networks, with IoT, the authentication of user become more challenging; however, a solution is presented in recent work.

Environmental/Earth Sensing

There are many applications in monitoring environmental parameters, examples of which are given below. They share the extra challenges of harsh environments and reduced power supply.

Air Pollution Monitoring

Wireless sensor networks have been deployed in several cities (Stockholm, London, and Brisbane) to monitor the concentration of dangerous gases for citizens. These can take advantage of the ad hoc wireless links rather than wired installations, which also make them more mobile for testing readings in different areas.

Forest Fire Detection

A network of Sensor Nodes can be installed in a forest to detect when a fire has started. The nodes can be equipped with sensors to measure temperature, humidity and gases which are produced by fire in the trees or vegetation. The early detection is crucial for a successful action of the firefighters; thanks to Wireless Sensor Networks, the fire brigade will be able to know when a fire is started and how it is spreading.

Landslide Detection

A landslide detection system makes use of a wireless sensor network to detect the slight movements of soil and changes in various parameters that may occur before or during a landslide. Through the data gathered it may be possible to know the impending occurrence of landslides long before it actually happens.

Water Quality Monitoring

Water quality monitoring involves analyzing water properties in dams, rivers, lakes and oceans, as well as underground water reserves. The use of many wireless distributed sensors enables the creation of a more accurate map of the water status, and allows the permanent deployment of monitoring stations in locations of difficult access, without the need of manual data retrieval.

Natural Disaster Prevention

Wireless sensor networks can effectively act to prevent the consequences of natural disasters, like floods. Wireless nodes have successfully been deployed in rivers where changes of the water levels have to be monitored in real time.

Industrial Monitoring

Machine Health Monitoring

Wireless sensor networks have been developed for machinery condition-based maintenance (CBM) as they offer significant cost savings and enable new functionality.

Wireless sensors can be placed in locations difficult or impossible to reach with a wired system, such as rotating machinery and untethered vehicles.

Data Center Monitoring

Due to the high density of servers racks in a data center, often cabling and IP addresses are an issue. To overcome that problem more and more racks are fitted out with wireless temperature sensors to monitor the intake and outtake temperatures of racks. As ASHRAE recommends up to 6 temperature sensors per rack, meshed wireless temperature technology gives an advantage compared to traditional cabled sensors.

Data Logging

Wireless sensor networks are also used for the collection of data for monitoring of environmental information, this can be as simple as the monitoring of the temperature in a fridge to the level of water in overflow tanks in nuclear power plants. The statistical information can then be used to show how systems have been working. The advantage of WSNs over conventional loggers is the "live" data feed that is possible.

Water/Waste Water Monitoring

Monitoring the quality and level of water includes many activities such as checking the quality of underground or surface water and ensuring a country's water infrastructure for the benefit of both human and animal. It may be used to protect the wastage of water.

Structural Health Monitoring

Wireless sensor networks can be used to monitor the condition of civil infrastructure and related geo-physical processes close to real time, and over long periods through data logging, using appropriately interfaced sensors.

Wine Production

Wireless sensor networks are used to monitor wine production, both in the field and the cellar.

Characteristics

The main characteristics of a WSN include

- Power consumption constraints for nodes using batteries or energy harvesting. Examples of suppliers are ReVibe Energy and Perpetuum

- Ability to cope with node failures (resilience)

- Some mobility of nodes (for highly mobile nodes see MWSNs)

- Heterogeneity of nodes

- Homogeneity of nodes

- Scalability to large scale of deployment

- Ability to withstand harsh environmental conditions

- Ease of use

- Cross-layer design

Cross-layer is becoming an important studying area for wireless communications. In addition, the traditional layered approach presents three main problems:

1. Traditional layered approach cannot share different information among different layers, which leads to each layer not having complete information. The traditional layered approach cannot guarantee the optimization of the entire network.

2. The traditional layered approach does not have the ability to adapt to the environmental change.

3. Because of the interference between the different users, access conflicts, fading, and the change of environment in the wireless sensor networks, traditional layered approach for wired networks is not applicable to wireless networks.

So the cross-layer can be used to make the optimal modulation to improve the transmission performance, such as data rate, energy efficiency, QoS (Quality of Service), etc. Sensor nodes can be imagined as small computers which are extremely basic in terms of their interfaces and their components. They usually consist of a *processing unit* with limited computational power and limited memory, *sensors* or MEMS (including specific conditioning circuitry), a *communication device*(usually radio transceivers or alternatively optical), and a power source usually in the form of a battery. Other possible inclusions are energy harvesting modules,secondary ASICs, and possibly secondary communication interface (e.g. RS-232 or USB).

The base stations are one or more components of the WSN with much more computational, energy and communication resources. They act as a gateway between sensor nodes and the end user as they typically forward data from the WSN on to a server.

Other special components in routing based networks are routers, designed to compute, calculate and distribute the routing tables.

Platforms

Hardware

One major challenge in a WSN is to produce *low cost* and *tiny* sensor nodes. There are an increasing number of small companies producing WSN hardware and the commercial situation can be compared to home computing in the 1970s. Many of the nodes are still in the research and development stage, particularly their software. Also inherent to sensor network adoption is the use of very low power methods for radio communication and data acquisition.

In many applications, a WSN communicates with a Local Area Network or Wide Area Network through a gateway. The Gateway acts as a bridge between the WSN and the other network. This enables data to be stored and processed by devices with more resources, for example, in a remotely located server. A wireless wide area network used primarily for low-power devices is known as a Low-Power Wide-Area Network (LPWAN).

Wireless

There are several wireless standards and solutions for sensor node connectivity. Thread and ZigBee can connect sensors operating at 2.4 GHz with a data rate of 250kbit/s. Many use a lower frequency to increase radio range (typically 1km), for example Z-wave operates at 915 MHz and in the EU 868 MHz has been widely used but these have a lower data rate (typically 50 kb/s). The IEEE 802.15.4 working group provides a standard for low power device connectivity and commonly sensors and smart meters use one of these standards for connectivity. With the emergence of Internet of Things, many other proposals have been made to provide sensor connectivity. LORA is a form of LPWAN which provides long range low power wireless connectivity for devices, which has been used in smart meters. Wi-SUN connects devices at home. NarrowBand IOT and LTE-M can connect up to millions of sensors and devices using cellular technology.

Software

Energy is the scarcest resource of WSN nodes, and it determines the lifetime of WSNs. WSNs may be deployed in large numbers in various environments, including remote and hostile regions, where ad hoc communications are a key component. For this reason, algorithms and protocols need to address the following issues:

- Increased lifespan
- Robustness and fault tolerance
- Self-configuration

Lifetime maximization: Energy/Power Consumption of the sensing device should be minimized and sensor nodes should be energy efficient since their limited energy resource determines their lifetime. To conserve power, wireless sensor nodes normally power off both the radio transmitter and the radio receiver when not in use.

Operating Systems

Operating systems for wireless sensor network nodes are typically less complex than general-purpose operating systems. They more strongly resemble embedded systems, for two reasons. First, wireless sensor networks are typically deployed with a particular application in mind, rather than as a general platform. Second, a need for low costs and low power leads most wireless sensor nodes to have low-power microcontrollers ensuring that mechanisms such as virtual memory are either unnecessary or too expensive to implement.

It is therefore possible to use embedded operating systems such as eCos or uC/OSfor sensor networks. However, such operating systems are often designed with real-time properties.

TinyOS is perhaps the first operating system specifically designed for wireless sensor networks. TinyOS is based on an event-driven programming model instead of multithreading. TinyOS programs are composed of *event handlers* and *tasks* with run-to-completion semantics. When an external event occurs, such as an incoming data packet or a sensor reading, TinyOS signals the appropriate event handler to handle the event. Event handlers can post tasks that are scheduled by the TinyOS kernel some time later.

LiteOS is a newly developed OS for wireless sensor networks, which provides UNIX-like abstraction and support for the C programming language.

Contiki is an OS which uses a simpler programming style in C while providing advances such as 6LoWPAN and Protothreads.

RIOT_(operating_system) is a more recent real-time OS including similar functionality to Contiki.

PreonVM is an OS for wireless sensor networks, which provides 6LoWPANbased on Contiki and support for the Java programming language.

Online Collaborative Sensor Data Management Platforms

Online collaborative sensor data management platforms are on-line database services that allow sensor owners to register and connect their devices to feed data into an on-line database for storage and also allow developers to connect to the database and build their own applications based on that data. Examples include Xively and the Wikisensing platform. Such platforms simplify online collaboration between users over diverse data sets ranging from energy and environment data to that collected from transport services. Other services include allowing developers to embed real-time graphs & widgets in websites; analyse and process historical data pulled from the data feeds; send real-time alerts from any datastream to control scripts, devices and environments.

The architecture of the Wikisensing system describes the key components of such systems to include APIs and interfaces for online collaborators, a middleware containing the business logic needed for the sensor data management and processing and a storage model suitable for the efficient storage and retrieval of large volumes of data.

Simulation

At present, agent-based modeling and simulation is the only paradigm which allows the simulation of complex behavior in the environments of wireless sensors (such as flocking). Agent-based simulation of wireless sensor and ad hoc networks is a relatively new paradigm. Agent-based modelling was originally based on social simulation.

Network simulators like Opnet, Tetcos NetSim and NS can be used to simulate a wireless sensor network.

Other Concepts

Security

Infrastructure-less architecture (i.e. no gateways are included, etc.) and inherent requirements (i.e. unattended working environment, etc.) of WSNs might pose several weak points that attract adversaries. Therefore, security is a big concern when WSNs are deployed for special applications such as military and healthcare. Owing to their unique characteristics, traditional security methods of computer networks would be useless (or less effective) for WSNs. Hence, lack of security mechanisms would cause intrusions towards those networks. These intrusions need to be detected and mitigation methods should be applied. More interested readers would refer to Butun *et al.*'s paper regarding intrusion detection systemsdevised for WSNs.

Distributed Sensor Network

If a centralized architecture is used in a sensor network and the central node fails, then the entire network will collapse, however the reliability of the sensor network can be increased by using a distributed control architecture. Distributed control is used in WSNs for the following reasons:

1. Sensor nodes are prone to failure,

2. For better collection of data,

3. To provide nodes with backup in case of failure of the central node.

There is also no centralised body to allocate the resources and they have to be self organized.

Data Integration and Sensor Web

The data gathered from wireless sensor networks is usually saved in the form of numerical data in a central base station. Additionally, the Open Geospatial Consortium (OGC) is specifying standards for interoperability interfaces and metadata encodings that enable real time integration of heterogeneous sensor webs into the Internet, allowing any individual to monitor or control wireless sensor networks through a web browser.

In-network Processing

To reduce communication costs some algorithms remove or reduce nodes' redundant sensor information and avoid forwarding data that is of no use. As nodes can inspect the data they forward, they can measure averages or directionality for example of readings from other nodes. For example, in sensing and monitoring applications, it is generally the case that neighboring sensor nodes monitoring an environmental feature typically register similar values. This kind of data redundancy due to the spatial correlation between sensor observations inspires techniques for in-network data aggregation and mining. Aggregation reduces the amount of network traffic which helps to reduce energy consumption on sensor nodes.Recently, it has been found that network gateways also play an important role in improving energy efficiency of sensor nodes by scheduling more resources for the nodes with more critical energy efficiency need and advanced energy efficient scheduling algorithms need to be implemented at network gateways for the improvement of the overall network energy efficiency.

Secure Data Aggregation

This is a form of in-network processing where sensor nodes are assumed to be unsecured with limited available energy, while the base station is assumed to be secure with unlimited available energy. Aggregation complicates the already existing security challenges for wireless sensor networks and requires new security techniques tailored specifically for this scenario. Providing security to aggregate data in wireless sensor networks is known as *secure data aggregation in WSN*. were the first few works discussing techniques for secure data aggregation in wireless sensor networks.

Two main security challenges in secure data aggregation are confidentiality and integrity of data. While encryption is traditionally used to provide end to end confidentiality in wireless sensor network, the aggregators in a secure data aggregation scenario need to decrypt the encrypted data to perform aggregation. This exposes the plaintext at the aggregators, making the data vulnerable to attacks from an adversary. Similarly an aggregator can inject false data into the aggregate and make the base station accept false data. Thus, while data aggregation improves energy efficiency of a network, it complicates the existing security challenges.

Limitations of Wireless Sensor Networks

1. Possess very little storage capacity – a few hundred kilobytes

2. Possess modest processing power-8MHz

3. Works in short communication range – consumes a lot of power

4. Requires minimal energy – constrains protocols

5. Have batteries with a finite life time

6. Passive devices provide little energy

Mobile Wireless Sensor Network

Mobile wireless sensor networks (MWSNs) have recently launched a growing popular class of WSN in which mobility plays a key role in the execution of the application. In recent years, mobility has become an important area of research for the WSN community. The increasing capabilities and the decreasing costs of mobile sensors make mobile sensor networks possible and practical. Although WSN deployments were never envisioned to be fully static, mobility was initially regarded as having several challenges that needed to be overcome, including connectivity, coverage, and energy consumption, among others.However, recent studies have been showing mobility in a more favorable light.

Mobile wireless sensor networks (MWSNs) play a vital role in today's real world applications in which the sensor nodes are mobile. MWSNs are much more versatile than static WSNs as the sensor nodes can be deployed in any scenario and cope with rapid topology changes. Mobile sensor nodes consist of a microcontroller, various sensors (i.e., light, temperature, humidity, pressure, mobility, etc.), a radio transceiver, and that is powered by a battery . The mobility models to define the movements towards/ away the sensor nodes, and how the mobile sensor nodes location, velocity and acceleration change over time, also predicts the future node positions.

Types of WSNs

Usually, the sensor nodes are deployed on land, underground and under water environments and that forms a WSN. Based on the sensor nodes deployment, a sensor network faces different challenges and constraints. Types of the WSNs are terrestrial, multimedia, underground, multi-media and mobile WSNs. In this chapter, we are discussing the overview of the mobile WSNs. According to the resources of the sensor nodes on an MWSN, it can be classified into homogeneous and heterogeneous MWSNs. Homogeneous MWSN consists of identical mobile sensor nodes and they may have unique properties. But, heterogeneous MWSN consists of a number of mobile sensor nodes with different abilities in node property such as battery power, memory size, computing power, sensing range, transmission range, and mobility, etc. Also, the nodes deployment of heterogeneous MWSN is more complex than homogeneous MWSN.

Reasons why Mobile Nodes are Considered in WSNs

Kay Romer and Friedemann Mattern investigated the design space of the wireless sensor networks and suggested many applications such as bird observation on great duck island, zebranet, cattle herding, bathymetry, glacier monitoring, cold chain management, ocean water monitoring, grape monitoring, power monitoring, rescue of avalanche victims, vital sign monitoring, tracking military vehicles, parts assembly, self-healing mine field, and sniper localization. Among 15 different applications, 10 applications are purely mobile and one of them is partially mobile. Therefore, mobile sensor nodes play an important role in humans real world applications.

Mobile Sensor Node Architecture

Usually, the sensor nodes are designed with one or more sensors (i.e., temperature, light, humidity, moisture, pressure, luminosity, proximity, etc.), microcontroller, external memory, radio transceiver, analog to digital converter (ADC), antenna and battery. Again, the nodes are limited on-board storage, battery power, processing and radio capacity due to their small size . However, the mobile sensor node architecture is almost similar to the normal sensor node. But, some additional units are considered on mobile sensor nodes such as localization/position finders, mobilizer, and power generator. The architecture of the mobile sensor node is shown in figure below. The location or position finder unit is used to identify the position of the sensor node and the mobilizer provides mobility for a sensor node. The power generator unit is responsible to generate a power for fulfilling further energy requirements of the sensor node by applying any specific techniques such as the solar cell.

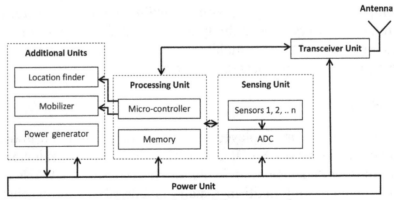

Architecture of the mobile sensor node.

Challenges

Broadly speaking, there are two sets of challenges in MWSNs; hardware and environment. The main hardware constraints are limited battery power and low cost requirements. The limited power means that it's important for the nodes to be energy efficient. Price limitations often demand low complexity algorithms for simpler microcontrollers and use of only a simplex radio. The major environmental factors are the shared medium and varying topology. The shared medium dictates that channel access must be regulated in some way. This is often done using a medium access control (MAC) scheme, such as carrier sense multiple access (CSMA), frequency division multiple access (FDMA) or code division multiple access (CDMA). The varying topology of the network comes from the mobility of nodes, which means that multihop paths from the sensors to the sink are not stable.

Standards

Currently there is no standard for MWSNs, so often protocols from MANETs are borrowed, such as Associativity-Based Routing (AR), Ad hoc On-Demand Distance Vector Routing (AODV), Dynamic Source Routing (DSR) and Greedy Perimeter Stateless Routing (GPSR). MANET protocols are preferred as they are able to work in mobile environments, whereas WSN protocols often aren't suitable.

Topology

Topology selection plays an important role in routing because the network topology decides the transmission path of the data packets to reach the proper destination. Here, all the topologies (Flat / Unstructured, cluster, tree, chain and hybrid topology) are not feasible for reliable data transmission on sensor nodes mobility. Instead of single topology, hybrid topology plays a vital role in data collection, and the performance is good. Hybrid topology management schemes include the Cluster Independent Data Collection Tree (CIDT). and the Velocity Energy-efficient and Link-aware Cluster-Tree (VELCT); both have been proposed for mobile wireless sensor networks (MWSNs).

Routing

Since there is no fixed topology in these networks, one of the greatest challenges is routing data from its source to the destination. Generally these routing protocols draw inspiration from two fields; WSNs and mobile ad hoc networks (MANETs). WSN routing protocols provide the required functionality but cannot handle the high frequency of topology changes. Whereas, MANET routing protocols can deal with mobility in the network but they are designed for two way communication, which in sensor networks is often not required.

Protocols designed specifically for MWSNs are almost always multihop and sometimes adaptations of existing protocols. For example, Angle-based Dynamic Source Routing (ADSR), is an adaptation of the wireless mesh network protocol Dynamic Source Routing (DSR) for MWSNs. ADSR uses location information to work out the angle between the node intending to transmit, potential forwarding nodes and the sink. This is then used to insure that packets are always forwarded towards the sink. Also, Low Energy Adaptive Clustering Hierarchy (LEACH) protocol for WSNs has been adapted to LEACH-M (LEACH-Mobile), for MWSNs. The main issue with hierarchical protocols is that mobile nodes are prone to frequently switching between clusters, which can cause large amounts of overhead from the nodes having to regularly re-associate themselves with different cluster heads.

Another popular routing technique is to utilise location information from a GPS module attached to the nodes. This can be seen in protocols such as Zone Based Routing (ZBR), which defines clusters geographically and uses the location information to keep nodes updated with the cluster they're in. In comparison, Geographically Opportunistic Routing (GOR), is a flat protocol that divides the network area into grids and then uses the location information to opportunistically forward data as far as possible in each hop.

Multipath protocols provide a robust mechanism for routing and therefore seem like a promising direction for MWSN routing protocols. One such protocol is the query based Data Centric Braided Multipath (DCBM).

Furthermore, Robust Ad-hoc Sensor Routing (RASeR) and Location Aware Sensor Routing (LASeR) are two protocols that are designed specifically for high speed MWSN applications, such as those that incorporate UAVs. They both take advantage of multipath routing, which is facilitated by a 'blind forwarding' technique. Blind forwarding simply allows the transmitting node to broadcast a packet to its neighbors, it is then the responsibility of the receiving nodes to decide whether they should forward the packet or drop it. The decision of whether to forward a packet or not is made using a network-wide gradient metric, such that the values of the transmitting and receiving nodes are compared to determine which is closer to the sink. The key difference between RASeR and LASeR is in the way they maintain their gradient metrics; RASeR uses the regular transmission of small beacon packets, in which nodes broadcast their current gradient. Whereas, LASeR relies on taking advantage of geographical location information that is already present on the mobile sensor node, which is likely the case in many applications.

Medium Access Control

There are three types of medium access control (MAC) techniques: based on time division, frequency division and code division. Due to the relative ease of implementation, the most common choice of MAC is time-division-based, closely related to the popular

CSMA/CA MAC. The vast majority of MAC protocols that have been designed with MWSNs in mind, are adapted from existing WSN MACs and focus on low power consumption, duty-cycled schemes.

Validation

Protocols designed for MWSNs are usually validated with the use of either analytical, simulation or experimental results. Detailed analytical results are mathematical in nature and can provide good approximations of protocol behaviour. Simulations can be performed using software such as OPNET, NetSim and NS2 and is the most common method of validation. Simulations can provide close approximations to the real behaviour of a protocol under various scenarios. Physical experiments are the most expensive to perform and, unlike the other two methods, no assumptions need to be made. This makes them the most reliable form of information, when determining how a protocol will perform under certain conditions.

Applications

The advantage of allowing the sensors to be mobile increases the number of applications beyond those for which static WSNs are used. Sensors can be attached to a number of platforms:

- People
- Animals
- Autonomous Vehicles
- Unmanned Vehicles
- Manned Vehicles

In order to characterise the requirements of an application, it can be categorised as either constant monitoring, event monitoring, constant mapping or event mapping. Constant type applications are time-based and as such data is generated periodically, whereas event type applications are event drive and so data is only generated when an event occurs. The monitoring applications are constantly running over a period of time, whereas mapping applications are usually deployed once in order to assess the current state of a phenomenon. Examples of applications include health monitoring, which may include heart rate, blood pressure etc. This can be constant, in the case of a patient in a hospital, or event driven in the case of a wearable sensor that automatically reports your location to an ambulance team in the case of an emergency. Animals can have sensors attached to them in order to track their movements for migration patterns, feeding habits or other research purposes. Sensors may also be attached to unmanned aerial vehicles (UAVs) for surveillance or environment mapping. In the case of autonomous UAV aided search and rescue, this would be considered an event mapping application, since the UAVs are deployed to search an area but will only transmit data back when a person has been found.

Sensor Grid

Recent advances in electronic circuit miniaturization and micro-electromechanical systems (MEMS) have led to the creation of small sensor nodes which integrate several kinds of sensors, a central processing unit (CPU), memory and a wireless transceiver. A collection of these sensor nodes forms a sensor network which is easily deployable to provide a high degree of visibility into real-world physical processes as they happen, thus benefitting a variety of applications such as environmental monitoring, surveillance and target tracking. Some of these sensor nodes may also incorporate actuators such as buzzers and switches which can affect the environment directly. We shall simply use the generic term sensor node to refer to these sensor-actuator nodes as well.

A parallel development in the technology landscape is grid computing, which is essentially the federation of heterogeneous computational servers connected by high-speed network connections. Middleware technologies such as Globus and Gridbus enable secure and convenient sharing of resources such as CPU, memory, storage, content and databases by users and applications. This has caused grid computing to be referred to as 'computing on tap', utility computing and IBM's mantra, 'on demand' computing. Many countries have recognized the importance of grid computing for 'eScience' and the grid has a number of success stories from the fields of bioinformatics, drug design, engineering design, business, manufacturing and logistics.

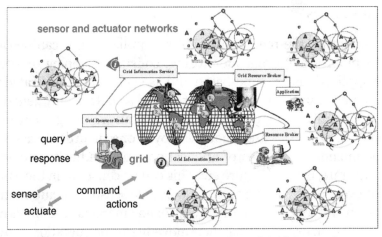

SensorGrid integrating Sensor Networks and Grid Computing.

The combination of sensor networks and grid computing in sensor-grid computing executing on a sensor-grid architecture (or simply a 'sensor-grid' in short) enables the complementary strengths and characteristics of sensor networks and grid computing to be realized on a single integrated platform. Essentially, sensor-grid computing combines real-time data about the environment with vast computational resources. This enables the construction of real-time models and databases of the environment and physical processes as they unfold, from which high-value computations like decision-making, analytics, data mining, optimization and prediction can be carried out to generate 'on-the-fly' results. This powerful combination would enable, for example, effective early warning of natural disasters such as tornados and tsunamis, and real-time business process optimization.

Approaches to Sensor-Grid Computing

One simple way to achieve sensor-grid computing is to simply connect and interface sensors and sensor networks to the grid and let all computations take place there. The grid will then issue commands to the appropriate actuators. In this case, all that is needed are high-speed communication links between the sensoractuator nodes and the grid. We refer to this as the centralized sensor-grid computing approach executing on acentralized sensor-grid architecture.

However, the centralized approach has several serious drawbacks. Firstly, it leads to excessive communications in the sensor network which rapidly depletes the batteries. It also does not take advantage of the in-network processing capability of sensor networks which permits simple processing and decision-making to be carried out close to the source of the sensed data. In the event of communication failure, such as when wireless communication in the sensor network is unavailable, e.g. due to jamming, the entire system becomes inoperational.

The more robust and efficient alternative is the decentralized or distributed sensor-grid computing approach which executes on a distributed sensor-grid architecture and alleviates most of the drawbacks of the centralized approach. The distributed sensor-grid computing approach involves processing and decision-making within the sensor network and at other levels of the sensor-grid architecture.

Concept and History

The concept of a sensor grid was first defined in the Discovery Net project where a distinction was made between "sensor networks" and "sensor grids".

Briefly whereas the design of a sensor network addresses the logical and physical connectivity of the sensors, the focus of constructing a sensor grid is on the issues relating to the data management, computation management, information management and knowledge discovery management associated with the sensors and the data they generate, and how they can be addressed within an open computing environment. In particular in a Sensor Grid is characterized by:

- Distributed Sensor Data Access and Integration: relating to both the heterogeneity and geographic distribution of the sensors within a sensor grid and how sensors can be located, accessed and integrated within a particular study.

- Large Data Set Storage and Management: relating to the sizes of data being collected and analyzed by multiple users at different locations for different purposes.

- Distributed Reference Data Access and Integration: relating to the need for integrating the analysis data collected from a Sensor Grid with other forms of data available of the Internet.

- Intensive and Open Data Analysis Computation: relating to the need for using a multitude of analysis components such as statistical, clustering, visualization and data classification tools that could be executing remotely on high performance computing servers on a computational Grid.

Uses

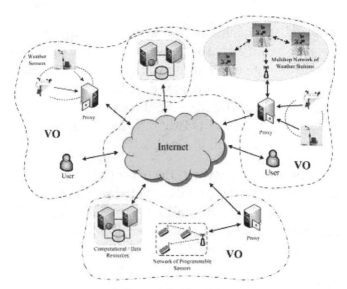

Typical sensor grid architecture

The sensor grid enables the collection, processing, sharing, visualization, archiving and searching of large amounts of sensor data. There are several rationales for a sensor grid. First, the vast amount of data collected by the sensors can be processed, analyzed, and stored using the computational and data storage resources of the grid. Second, the sensors can be efficiently shared by different users and applications under flexible usage scenarios. Each user can access a subset of the sensors during a particular time period to run a specific application, and to collect the desired type of sensor data. Third, as sensor devices with embedded processors become more computationally powerful, it is more efficient to offload specialized tasks such as image and signal on the sensor devices. Finally, a sensor grid provides seamless access to a wide variety of resources in a pervasive manner. Advanced techniques in artificial intelligence, data fusion, data mining, and distributed database processing can be applied to make sense of the sensor data and generate new knowledge of the environment. The results can in turn be used to optimize the operation of the sensors, or influence the operation of actuators to change the environment. Thus, sensor grids are well suited for adaptive and pervasive computing applications.

Applications

A sensor grid based architecture has many applications such as environmental and habitat monitoring, healthcare monitoring of patients, weather monitoring and forecasting, military and homeland security surveillance, tracking of goods and manufacturing processes, safety monitoring of physical structures and construction sites, smart homes and offices, and many other uses currently beyond our imagination. Various architectures that can be used for such applications as well as different kinds of data analysis and data mining that can be conducted.

Implementations of Distributed Sensor-Grid Computing

Distributed information fusion and distributed decision-making are two applications that are well-suited for distributed sensor-grid computing.

Distributed Information Fusion

Since the nodes in a sensor network are independently sensing the environment, this gives rise to a high degree of redundant information. However, due to the severely resource-constrained nature of sensor nodes, some of these readings may be inaccurate. Information fusion algorithms compute the most probable sensor readings and have been studied extensively over the years in the context of target detection and tracking.

We implemented a hierarchical decision fusion system comprising two levels of Crossbow motes (at the local or ground level), grid clients (at the regional level) and grid server nodes (at the global level) to detect and classify forest fires of varying degrees of severity, ranging from 'local fire', 'small forest fire' to 'large forest fire'. The local classifier in each sensor node is a Bayesian Maximum A Posteriori (MAP) classifier and the decision fusion algorithm described in Duarte and Hu is implemented at the fusion centers. During operation, the decision fusion algorithm produces the final classification outcome based on the most frequent class label among the training samples which produced the same decision vector as the one encountered during operation. This decision fusion algorithm is robust and produces high classification accuracy in the final classification even in the presence of faulty or noisy sensors.

In an enhanced version of the above, two further levels in the form of Stargate and iPAQ cluster heads were added between the sensor nodes and grid client levels. The resulting system can be seen in. The addition of these two levels enable more complex processing to be done close to the source of the sensor data and reduces the communication distances between the different levels, thus conserving power and improving the timeliness of the global classification.

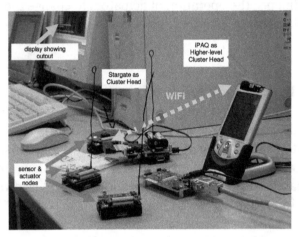

Hierarchical decision fusion system on sensor-grid architecture.

Distributed autonomous decision-making

There are many cases in which some response is needed from the sensor-grid system, but the best action to take in different situations or states is not known in advance. This can be determined through an adaptive learning process, such as the Markov Decision Process (MDP) or reinforcement learning (RL) approach. MDP problems can be solved off-line using methods such as policy- or value-iteration, or on-line using RL or neuro-dynamic programming (NDP) methods.

A multi-level distributed autonomous decision-making system can be implemented on the hierarchical sensorgrid architecture shown in. We implemented basic NDP agents in Crossbow motes at the local or ground level, and more complex NDP agents at grid server nodes at the core of the grid. Each NDP agent is able to act autonomously such that the entire sensor-grid remains responsive despite communication failures due to radio jamming, router faults etc.

Sensor Node

The wireless sensor nodes are the central elements in a wireless sensor network (WSN). The node consists of sensing, processing, communication, and power subsystems. The processor subsystem is the central element of the node and the choice of a processor determines the trade off between flexibility and efficiency - in terms of both energy and performance. There are several processors as options: microcontrollers, digital signal processors, application-specific integrated circuits, and field programmable gate arrays. The communication subsystem can be interfaced with the processor subsystem in different ways. The communication subsystem is the most energy intensive subsystem and its power consumption should be regulated.

Sensor Node Architecture

A sensor is a transducer which measures a physical quantity such as light, temperature, pressure and humidity. Sensor converts physical quantities into a electrical signal which can be used by a human or by an instrument to take necessary decisions. They self-organize and collaboratively coordinate the sensing process depending on the phenomenon. Sensors are typically small, battery-powered, low cost devices with wireless communication capability. A sensor node in a sensor network capable of performing processing, gathering sensory information and communicating with other connected nodes in the network. A general hardware architecture of a sensor node platform is shown in. Basically a typical sensor node made up of following four basic components: processing unit, communication unit, sensing unit and power unit. In addition to these, a sensor node may also contain some application specific components: mobilizer, location finding system and power generator.

Processing Unit

In a sensor node, the functionality of processing unit is to convert the electrical signals received from the sensor into an intelligible message format, schedule tasks, process data received, execute the algorithms for data forwarding and control the functionality of other hardware components in the sensor node. The processing unit made up of embedded processor which include Microcontroller, Digital Signal Processor (DSP), Field Programmable Gate Array (FPGA) and Application Specific Integrated Circuit (ASIC).

Communication Unit

The communication unit has both a transmitter and a receiver for establishing wireless communication between sensor nodes. The communication unit which combines both transmitting and receiving tasks are called transceiver. The most essential task of transceiver is to convert the digital bit stream coming from microcontroller into radio waves and vice versa. The Radio Frequency (RF) based wireless communication suits to most of WSN applications. Transmit, Receive, Idle and Sleep are the operational states of transceiver. The commercially available transceiver incorporates all the circuitry required for modulation, demodulation, amplification, filtering, mixing and so on.

Some of the standard radio transceivers used in various sensor nodes include RFM TR1000 family from RF Monolithics, CC1000 and CC2420 family from Chipcon, TDA 525x family from Infineon, IEEE 802.15.4, LMX3162 from National Semiconductor, RDSSS9M from Conexant systems.

Sensing Unit

Sensing unit consist of two subunits: sensors and analog to digital converters (ADCs). Sensor is a transducer which produces a measurable electrical signal to a change in a physical phenomena such as temperature, chemical level, light intensity, sound, magnetic fields, image, etc. Sensors can be classified as either analog or digital devices. The sensor produces continuous analog signals, converted into digital signals by ADCs and then fed to the processor for further processing.

Power Unit

The electronics of the sensor node is powered by using either stored energy or harvesting energy from other potential sources such as light, vibration, heat and radio frequency signal. Energy supply is necessary to support network operation from a few hours to months or even years. Most of the existing commercial and research platforms rely on batteries, which dominate the node size. Batteries are the obvious energy storage medium, ranging from the small coin cell to large sealed lead acid batteries (AA, AAA types) which include lead acid, lithium, NiCad, NiMH, and thin-film. Rechargeable batteries are typically not desirable due to lower energy density, higher cost and in most of the applications recharging is simply impractical.

Application Specific Units

In addition to the basic units, sensor nodes might have some additional application specific units such as location finding systems, mobilizer and power generator. In most of the application, WSN routing techniques and sensing tasks require the information of location. So location finding systems or Global Positioning System (GPS) are used to estimate the geographical position with certain level of accuracy. Some applications require the sensor node to move to carry out the assigned task. In such cases the sensor nodes equipped with mobility unit called mobilizer. The power generator which supplies continuous power to a node have some additional functionality.

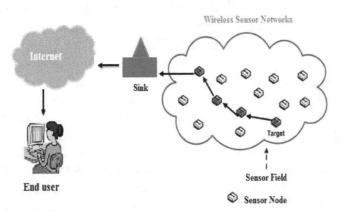

Typical Wireless Sensor Network Architecture

Components

The main components of a sensor node are a microcontroller, transceiver, external memory, power source and one or more sensors.

Controller

The controller performs tasks, processes data and controls the functionality of other components in the sensor node. While the most common controller is a microcontroller, other alternatives that can be used as a controller are: a general purpose desktop microprocessor, digital signal processors, FPGAs and ASICs. A microcontroller is often used in many embedded systems such as sensor nodes because of its low cost, flexibility to connect to other devices, ease of programming, and low power consumption. A general purpose microprocessor generally has a higher power consumption than a microcontroller, therefore it is often not considered a suitable choice for a sensor node. Digital Signal Processors may be chosen for broadband wireless communication applications, but in Wireless Sensor Networks the wireless communication is often modest: i.e., simpler, easier to process modulation and the signal processing tasks of actual sensing of data is less complicated. Therefore, the advantages of DSPs are not usually of much importance to wireless sensor nodes. FPGAs can be reprogrammed and reconfigured according to requirements, but this takes more time and energy than desired.

Transceiver

Sensor nodes often make use of ISM band, which gives free radio, spectrum allocation and global availability. The possible choices of wireless transmission media are radio frequency (RF), optical communication (laser) and infrared. Lasers require less energy , but need line-of-sight for communication and are sensitive to atmospheric conditions. Infrared, like lasers, needs no antenna but it is limited in its broadcasting capacity. Radio frequency-based communication is the most relevant that fits most of the WSN applications. WSNs tend to use license-free communication frequencies: 173, 433, 868, and 915 MHz; and 2.4 GHz. The functionality of both transmitter and receiver are combined into a single device known as a transceiver. Transceivers often lack unique identifiers. The operational states are transmit, receive, idle, and sleep. Current generation transceivers have built-in state machines that perform some operations automatically.

Most transceivers operating in idle mode have a power consumption almost equal to the power consumed in receive mode. Thus, it is better to completely shut down the transceiver rather than leave it in the idle mode when it is not transmitting or receiving. A significant amount of power is consumed when switching from sleep mode to transmit mode in order to transmit a packet.

External Memory

From an energy perspective, the most relevant kinds of memory are the on-chip memory of a microcontroller and Flash memory—off-chip RAM is rarely, if ever, used. Flash memories are used due to their cost and storage capacity. Memory requirements are very much application dependent. Two categories of memory based on the purpose of storage are: user memory used for storing application related or personal data, and program memory used for programming the device. Program memory also contains identification data of the device if present.

Power Source

A wireless sensor node is a popular solution when it is difficult or impossible to run a mains supply to the sensor node. However, since the wireless sensor node is often placed in a hard-to-reach location, changing the battery regularly can be costly and inconvenient. An important aspect in the development of a wireless sensor node is ensuring that there is always adequate energy available to power the system. The sensor node consumes power for sensing, communicating and data processing. More energy is required for data communication than any other process. The energy cost of transmitting 1 Kb a distance of 100 metres (330 ft) is approximately the same as that used for the execution of 3 million instructions by a 100 million instructions per second/W processor. Power is stored either in batteries or capacitors. Batteries, both rechargeable and non-rechargeable, are the main source of power supply for sensor nodes. They are also classified according to electrochemical material used for the electrodes such as NiCd (nickel-cadmium), NiZn (nickel-zinc), NiMH (nickel-metal hydride), and lithium-ion. Current sensors are able to renew their energy from solar sources, temperature differences, or vibration. Two power saving policies used are Dynamic Power Management (DPM) and Dynamic Voltage Scaling (DVS). DPM conserves power by shutting down parts of the sensor node which are not currently used or active. A DVS scheme varies the power levels within the sensor node depending on the non-deterministic workload. By varying the voltage along with the frequency, it is possible to obtain quadratic reduction in power consumption.

Sensors

Sensors are used by wireless sensor nodes to capture data from their environment. They are hardware devices that produce a measurable response to a change in a physical condition like temperature or pressure. Sensors measure physical data of the parameter to be monitored and have specific characteristics such as accuracy, sensitivity etc. The continual analog signal produced by the sensors is digitized by an analog-to-digital converter and sent to controllers for further processing. Some sensors contain the necessary electronics to convert the raw signals into readings which can be retrieved via a digital link (e.g. I2C, SPI) and many convert to units such as °C. Most sensor nodes are small in size, consume little energy, operate in high volumetric densities, be autonomous and operate unattended, and be adaptive to the environment. As wireless sensor nodes are typically very small electronic devices, they can only be equipped with a limited power source of less than 0.5-2 ampere-hour and 1.2-3.7 volts.

Sensors are classified into three categories: passive, omnidirectional sensors; passive, narrow-beam sensors; and active sensors. Passive sensors sense the data without actually manipulating the environment by active probing. They are self powered; that is, energy is needed only to amplify their analog signal. Active sensors actively probe the environment, for example, a sonar or radar sensor, and they require continuous energy from a power source. Narrow-beam sensors have a well-defined notion of direction of measurement, similar to a camera. Omnidirectional sensors have no notion of direction involved in their measurements.

Most theoretical work on WSNs assumes the use of passive, omnidirectional sensors. Each sensor node has a certain area of coverage for which it can reliably and accurately report the particular quantity that it is observing. Several sources of power consumption in sensors are: signal sampling and conversion of physical signals to electrical ones, signal conditioning, and analog-to-digital conversion. Spatial density of sensor nodes in the field may be as high as 20 nodes per cubic meter.

Visual Sensor Network

Visual sensor networks have emerged as an important class of sensor-based distributed intelligent systems, with unique performance, complexity, and quality of service challenges. Consisting of a large number of low-power camera nodes, visual sensor networks support a great number of novel vision-based applications. The camera nodes provide information from a monitored site, performing distributed and collaborative processing of their collected data. Using multiple cameras in the network provides different views of the scene, which enhances the reliability of the captured events. However, the large amount of image data produced by the cameras combined with the network's resource constraints require exploring new means for data processing, communication, and sensor management. Meeting these challenges of visual sensor networks requires interdisciplinary approaches, utilizing vision processing, communications and networking, and embedded processing.

Camera-based networks have been used for security monitoring and surveillance for a very long time. In these networks, surveillance cameras act as independent peers that continuously send video streams to a central processing server, where the video is analyzed by a human operator.

With the advances in image sensor technology, low-power image sensors have appeared in a number of products, such as cell phones, toys, computers, and robots. Furthermore, recent developments in sensor networking and distributed processing have encouraged the use of image sensors in these networks, which has resulted in a new ubiquitous paradigm—visual sensor networks. Visual sensor networks (VSNs) consist of tiny visual sensor nodes called camera nodes, which integrate the image sensor, embedded processor, and wireless transceiver. Following the trends in low-power processing, wireless networking, and distributed sensing, visual sensor networks have developed as a new technology with a number of potential applications, ranging from security to monitoring to telepresence.

In a visual sensor network a large number of camera nodes form a distributed system, where the camera nodes are able to process image data locally and to extract relevant information, to collaborate with other cameras on the application-specific task, and to provide the system's user with information-rich descriptions of captured events. With current trends moving toward development of distributed processing systems and with an increasing number of devices with built-in image sensors, a question of how these devices can be used together appears . There are several specific questions that have intrigued the research community. How can the knowledge gained from wireless sensor networks be used in the development of visual sensor networks? What kind of data processing algorithms can be supported by these networks? What is the best way to manage a large number of cameras in an efficient and scalable manner? What are the most efficient camera node architectures?

Characteristics of Visual Sensor Networks

One of the main differences between visual sensor networks and other types of sensor networks lies in the nature of how the image sensors perceive information from the environment. Most sensors provide measurements as 1D data signals. However, image sensors are composed of a large number of photosensitive cells. One measurement of the image sensor provides a 2D set of data points, which we see as an image. The additional dimensionality of the data set results in richer information content as well as in a higher complexity of data processing and analysis.

In addition, a camera's sensing model is inherently different from the sensing model of any other type of sensor. Typically, a sensor collects data from its vicinity, as determined by its sensing range. Cameras, on the other hand, are characterized by a directional sensing model—cameras capture images of distant objects/scenes from a certain direction. The 2D sensing range of traditional sensor nodes is, in the case of cameras, replaced by a 3D viewing volume (called field of view, or FoV).

Visual sensor networks are in many ways unique and more challenging compared to other types of wireless sensor networks. These unique characteristics of visual sensor networks are described next.

Resource Requirements

The lifetime of battery-operated camera nodes is limited by their energy consumption, which is proportional to the energy required for sensing, processing, and transmitting the data. Given the large amount of data generated by the camera nodes, both processing and transmitting image data are quite costly in terms of energy, much more so than for other types of sensor networks. Furthermore, visual sensor networks require large bandwidth for transmitting image data. Thus both energy and bandwidth are even more constrained than in other types of wireless sensor networks.

Local Processing

Local (on-board) processing of the image data reduces the total amount of data that needs to be communicated through the network. Local processing can involve simple image processing algorithms (such as background substraction for motion/object detection, and edge detection) as well as more complex image/vision processing algorithms (such as feature extraction, object classification, scene reasoning). Thus, depending on the application, the camera nodes may provide different levels of intelligence, as determined by the complexity of the processing algorithms they use . For example, low-level processing algorithms (such as frame differencing for motion detection or edge detection algorithms) can provide a camera node with the basic information about the environment, and help it decide whether it is necessary to transmit the captured image or whether it should continue processing the image at a higher level. More complex vision algorithms (such as object feature extraction, object classification, etc.) enable cameras to reason about the captured phenomena, such as to provide basic classification of the captured object. Furthermore, the cameras can collaborate by exchanging the detected object features, enabling further processing to collectively reason about the object's appearance or behavior. At this point the visual sensor network becomes a user-independent, intelligent system of distributed cameras that provides only relevant information about the monitored phenomena. Therefore, the increased complexity of vision processing algorithms results in highly intelligent camera systems that are oftentimes called smart camera networks .

In order to extract necessary information from different images, a camera node must employ different image processing algorithms. One specific image processing algorithm cannot achieve the same performance for different types of images—for example, an algorithm for face extraction significantly differs from algorithm for vehicle detection. However, oftentimes it is impossible to keep all the necessary image processing algorithms in the constrained memory of a camera node. One solution to this problem is to use mobile agents—a specific piece of software dispatched by the sink node to the region of interest . Mobile agents collect and aggregate the data using a specific image algorithm and send the processed data back to the sink. Furthermore, the mobile agents can migrate between the nodes in order to perform the specific task, thereby performing distributed information processing . In this way, the amount of data sent by the node, as well as the number of data flows in the network, can be significantly reduced.

Real-Time Performance

Most applications of visual sensor networks require real-time data from the camera nodes, which imposes strict boundaries on maximum allowable delays of data from the sources (cameras) to the user (sink). The real-time performance of a visual sensor network is affected by the time required for image data processing and for the transmission of the processed data throughout the network. Constrained by limited energy resources and by the processing speed of embedded processors, most camera nodes have processors that support only lightweight processing algorithms. On the network side, the real-time performance of a visual sensor network is constrained by the wireless channel limitations (available bandwidth, modulation, data rate), employed wireless standard, and by the current network condition. For example, upon detection of an event, the camera nodes can suddenly inject large amounts of data in the network, which can cause data congestion and increase data delays.

Different error protection schemes can affect the real-time transmission of image data through the network as well. Commonly used error protection schemes, such as automated-repeat-request (ARQ) and forward-error-correction (FEC) have been investigated in order to increase the reliability of wireless data transmissions . However, due to the tight delay constraints, methods such as ARQ are not suitable to be used in visual sensor networks. On the other hand, FEC schemes usually require long blocks in order to perform well, which again can jeopardize delay constraints.

Finally, multihop routing is the preferred routing method in wireless sensor networks due to its energy-efficiency. However, multihop routing may result in increased delays, due to queueing and data processing at the intermediate nodes. Thus, the total delay from the data source (camera node) to the sink increases with the number of hops on the routing path. Additionally, bandwidth becomes a scarce resource in multihop networks consisting of traditional wireless sensor nodes. In order to support the transmission of real-time data, different wireless modules that provide larger bandwidths (such as those based on IEEE 802.11 b,g,n) can be considered.

Precise Location and Orientation Information

In visual sensor networks, most of the image processing algorithms require information about the locations of the camera nodes as well as information about the cameras' orientations. This information can be obtained through a camera calibration process, which retrieves information on the cameras' intrinsic and extrinsic parameters. Estimation of calibration parameters usually requires knowledge of a set of feature point correspondences among the images of the cameras. When this is not provided, the cameras can be calibrated up to a similarity transformation , meaning that only relative coordinates and orientations of the cameras with respect to each other can be determined.

Time Synchronization

The information content of an image may become meaningless without proper information about the time at which this image was captured. Many processing tasks that involve multiple cameras (such as object localization) depend on highly synchronized cameras' snapshots. Time synchronization protocols developed for wireless sensor networks can be successfully used for synchronization of visual sensor networks as well.

Data Storage

The cameras generate large amounts of data over time, which in some cases should be stored for later analysis. An example is monitoring of remote areas by a group of camera nodes, where the frequent transmission of captured image data to a remote sink would quickly exhaust the cameras' energy resources. Thus, in these cases the camera nodes should be equipped with memories of larger capacity in order to store the data. To minimize the amount of data that requires storage, the camera node should classify the data according to its importance by using spatiotemporal analysis of image frames, and decide which data should have priority to be stored. For example, if an application is interested in information about some particular object, then the background can be highly compressed and stored, or even completely discarded .

The stored image data usually becomes less important over time, so it can be substituted with newly acquired data. In addition, reducing the redundancy in the data collected by cameras with overlapped views can be achieved via local communication and processing. This enables the cameras to reduce their needs for storage space by keeping only data of unique image regions. Finally, by increasing the available memory, more complex processing tasks can be supported on-board, which in return can reduce data transmissions and reduce the space needed for storing processed data.

Autonomous Camera Collaboration

Visual sensor networks are envisioned as distributed and autonomous systems, where cameras collaborate and, based on exchanged information, reason autonomously about the captured event and decide how to proceed. Through collaboration, the cameras relate the events captured in the images, and they enhance their understanding of the environment. Similar to wireless sensor networks, visual sensor networks should be data-centric, where captured events are described by their names and attributes. Communication between cameras should be based on some uniform ontology for the description of the event and interpretation of the scene dynamics.

Applications of Visual Sensor Networks

With the rapid development of visual sensor networks, numerous applications for these networks have been envisioned, as illustrated in the Table. Here, we mention some of these applications.

General application	Specific application
Surveillance	Public places
	Traffic
	Parking lots
	Remote areas
	Hazardous areas
Environmental monitoring	Animal habitats
	Building monitoring
Smart homes	Elderly care
	Kindergarten
Smart meeting rooms	Teleconferencing
	Virtual studios
Virtual reality	Telepresence systems
	Telereality systems

Table: Applications of visual sensor networks

(i) Surveillance: Surveillance has been the primary application of camera-based networks for a long time, where the monitoring of large public areas (such as airports, subways, etc.) is performed by hundreds or even thousands of security cameras. Since cameras usually provide raw video streams, acquiring important information from collected image data requires a huge amount of processing and human resources, making it time-consuming and prone to error. Current efforts in visual sensor networking are concentrated toward advancing the existing surveillance technology by utilizing intelligent methods for extracting information from image data locally on the camera node, thereby reducing the amount of data traffic. At the same time, visual sensor networks integrate resource-aware camera management policies and wireless networking aspects with surveillance-specific tasks. Thus, visual sensor networks can be seen as a next generation of surveillance systems that are not limited by the absence of infrastructure, nor do they require large processing resources at one central server. These networks are adaptable to the environment dynamics, autonomous, and able to respond timely to a user's requests by providing an immediate view from any desired viewpoint or by analyzing and providing information from specific, user determined areas.

(ii) Environmental monitoring: Visual sensor networks can be used for monitoring remote and inaccessible areas over a long period of time. In these applications, energy-efficient operations are particularly important in order to prolong monitoring over an extended period of time. Oftentimes the cameras are combined with other types of sensors into a heterogeneous network, such that the cameras are triggered only when an event is detected by other sensors used in the network .

(iii) Smart homes: There are situations (such as patients in hospitals or people with disabilities), where a person must be under the constant care of others. Visual sensor networks can provide continuous monitoring of people, and using smart algorithms the network can provide information about the person needing care, such as information about any unusual behavior or an emergency situation.

(iv) Smart meeting rooms: Remote participants in a meeting can enjoy a dynamic visual experience using visual and audio sensor network technology.

(v) Telepresence systems: Telepresence systems enable a remote user to "visit" some location that is monitored by a collection of cameras. For example, museums, galleries or exhibition rooms can be covered by a network of camera nodes that provide live video streams to a user who wishes to access the place remotely (e.g., over the Internet). The system is able to provide the user with any current view from any viewing point, and thus it provides the sense of being physically present at a remote location through interaction with the system's interface . Telereality aims to synthesize realistic novel views from images acquired from multiple cameras .

Virtual Sensor Network

Over the last few years, we are witnessing an explosion of interest in research on Wireless Sensor Networks (WSNs) that are accompanied by a constantly increasing interest for physical WSN deployment and commercial exploitation of the services provided by these networks. In this regard, the majority of relevant research and development efforts have focused on WSN systems that are dedicated for a specific application. However, a trend is currently developing towards resource-rich WSN deployments that are expected to provide capabilities in excess of any application's requirements.

In that sense, sensor-based applications/services will utilize sensors for purposes beyond the scope of the original sensor design and deployment, creating in this way an instantiation of a Virtual Sensor Network (VSN) [1,2] to serve user-specific requests. Virtualization is the key enabler for decoupling the physical sensor deployment from the applications running on top, and is thus a significant step towards the decoupling of ownerships in the Internet of Things. In rough analogy to virtualization in more "traditional" network infrastructures (both at the core and access network levels) , WSN virtualization aims to accommodate multiple logical network instances over a single physical network infrastructure with the ultimate goal of (a) supporting applications with different requirements both in terms of nodes and communication functionalities and (b) utilizing in an efficient and cost-effective manner the available network resources.

With these goals in mind, one could think of several applications that can benefit from the virtualization of sensor network infrastructures. In the following, we discuss two types of applications that correspond to what we consider to be the most significant use cases for VSNs. The first type involves geographically overlapped applications . In this case, a WSN that is deployed to support an application can utilize communication resources offered by another WSN operating in the same area and being deployed to support a different application. The main benefit from the collaboration of the different WSNs, in this case, is the reduction of the number of sensors of each type without losing accuracy or degrading the required user functionality. With the smart usage of previously deployed networks, this approach contributes to cost savings and helps future evolvement and addition of new types of sensors without the need to deploy a full network from scratch.

The second VSN use case involves applications operating over multi-purpose sensor networks. A common use of VSNs, in this case, could be the separation of nodes with multiple sensing elements in different (virtual) networks, each one being responsible to track a different sensing parameter. Using this approach, a better administration at node level and enhanced usability can be achieved with an increased number of different end-users gaining access and control in sensors' information according to their needs.

Whilst mechanisms are in place to facilitate virtualization in the core network , major challenges remain innetworks that connect embedded sensor devices to the Internet. On the one hand, these have to do with individual functionalities of a VSN system related, for example, to middleware, routing, security, trust, and energy awareness of protocols and mechanisms at various layers of the communication protocol stack. On the other hand, significant challenges arise as far as the development of a holistic approach to VSN is concerned, which should take into account not only individual enhancements related to the aforementioned functions, but also the need to federate resources across networks belonging to different administrative domains, as well as the need to support promising business models

Applications

VSNs are useful in three major classes of applications:

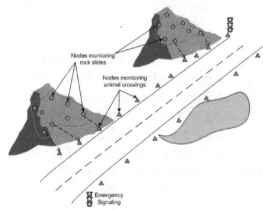

A single WSN is deployed to monitor rock sliding and animal crossing in a mountainous retain (two applications). Each application use nodes from the other application to relay its data to the signaling systemsand/or to its members.

- Geographically overlapped applications

  o E.g., monitoring rock slides and animal crossing within a mountainous terrain. Different types of devices that detect these phenomena can relay each other for data transfer without having to deploy separate networks. Here the advantage is saving in hardware cost.

- While logically separating multi-purpose sensor networks

  o E.g., smart neighborhood systems with multifunctional sensor nodes. Instead of traditional WSNs that runs one single applications, VSN enabled nodes run multiple applications

- In certain dedicated but dynamic applications

  o E.g., To enhance efficiency of a system that track dynamic phenomena such as subsurface chemical plumes that migrate, split, or merge. Such networks may involve dynamically varying subsets of sensors. Here the advantage is the ability to connect right set of nodes at the right time.

Examples

1. Example: Geographically overlapped sensing applications

Consider two sensor-network based systems, one for warning of rock slides and the other for warning of presence of wildlife, both to be deployed on a section of road passing through a mountainous terrain. With the traditional approach, these functions would be carried out by two separate sensor networks. Resource sharing between these two systems can result in significant efficiencies. By resource sharing, we do not mean that each node implement both the functionalities as that would effectively mean a dedicated sensor network with its functional specifications redefined to include both warning functions. Instead what we refer to here is the existence of two sensor networks in a symbiotic relationship providing one's resources to the other and using the resources of the other in such a way as to benefit both the systems.

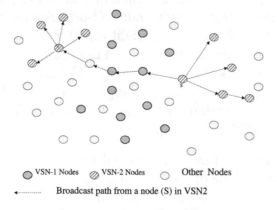

Broadcast path from a node (S) in VSN2

Consider as an example routing of messages in a sensor network. Sensor networks typically use multi-hop transmissions due to constraints on transmit power; it results in significant power savings and is often the only way to cover a large area. With two independent networks, the nodes in each network have to be deployed at a certain density; the nodes detecting rock-slides will have to be placed even in places where there are no rocks, and in fact power constraints may dictate where the nodes should be placed instead of application constraints. There are other reasons as well, such as fault tolerance, for these two networks that overlap partially in physical space to share resources to a limited degree, resulting in efficient resource usage and better performance attributes. However, this symbiotic relationship has to be achieved without sacrificing the advantages of individual dedicated sensor networks, especiallyn the relative ease of development of an application running over a dedicated set of nodes. With a VSN based implementation, certain nodes will be engaged in tracking wild life, others tracking rock slides, and yet others providing communication support. The VSNs membership of nodes may change with time and external events. If a third set of VSN capable sensor nodes for rainfall measurements is deployed in the same area, for example, these nodes become a part of the existing physical network, and use its resources for implementing the rainfall measurement related functionality while supporting other nodes in providing VSN support functions.

2. Example: Underground Contaminant Plume Tracking

Next we describe a sensor network dedicated to one application, yet one that benefits by the use of VSN concept. A chemical plume can be considered as a 3-D transient phenomenon that is spatially and temporally distributed, and which evolves in its intensity and extent. Hence, it is different from a phenomenon that is time varying in a fixed region (such as temperature/humidity changes in a room), or a phenomenon that varies in locations but not extents (such as a mobile object). For instance, plumes can change their configuration/shapes as a result of not only migration but also remedial treatment. In other words, two plumes can merge into one, and one plume can also be separated into two. As a result of this, a sensor node should adapt to plume dynamics and change its functionality to either active sensing (when they are embedded in plumes) or passive listening (when they are emerged out of plumes). This further implies that the sensor nodes should self-organize themselves to ensure that the right set of nodes collaborate at the right time for sensing and tracking a given plume, and the collected data can be delivered to the appropriate nodes, or perhaps even a remote server, for processing in an energy efficient manner. When two plumes merge, the corresponding VSNs will also merge to form one VSN tracking the newly formed plume. Similarly, breakup of a large plume into different plumes should result in the partition of VSN into multiple VSNs.

The sensor network for the three-dimensional transient plume is based on a two-tier sensor network consisting of a set of surface sensor nodes (S-nodes) and strings of sensors (W-nodes) placed within each well in a vertical array as shown in Figure. Each string array consists of sensors (Wnodes) placed at different depths measuring different variables (pressure, concentration, temperature, etc.). The node on the surface (S-node) at each well, in addition to monitoring,n serves as a computation and communication node on behalf of the corresponding W-nodes. The S nodes form a self-configuring wireless network. Each node is at a fixedb location, and the knowledge of its geographical location information is available for the application. W-nodes affected by a transient plume detect levels of concentration and pressure.

As Figure below indicates, the sensor nodes monitoring a plume are not necessarily adjacent in terms of connectivity. This subgroup and any other sensors of interest for tracking this plume (for example downstream nodes) are to be considered as a virtual sensor network. The shape and concentration profile of the plume as well as its migration path dictates the membership of the VSN. As the plume migrates,

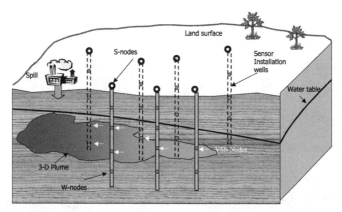

Figure: Physical Sensor Network Deployment for 3D transient plumes monitoring

The W-nodes within and on the boundary of the plume form a virtual sensor network (VSN). The remaining nodes provide VSN support services, such as routing among nodes in the VSN. The VSN will change with and in anticipation of plume movement.

the membership of VSN changes. The sensor network, by itself or based on prediction of models, alerts additional sensor nodes in the predicted and possible paths. The plume monitoring task thus will be based on the concept of Virtual Sensor Networks (VSNs), where multiple VSNs exist on a physical sensor network. While some of the nodes in a VSN may be able to communicate directly with each other, the VSN may consist of multiple zones, where communication between zones have to rely on nodes that are not members of the VSN. Thus providing communication support for maintaining the VSN is a key support function that is necessary.

The membership in VSN is dynamic, and the communications among VSN nodes frequently rely on other nodes. Functionality expected of a node will be dependent on whether or not it is currently a member of a VSN. Those that are not in the VSN need to support the maintenance of VSNs, while those within the VSN need to carry out tasks such as profile detection, pattern recognition, and tracking.

VSN Support Functions

The major functions of VSN can be divided into two categories: VSN maintenance and membership maintenance. The membership in VSN is dynamic, and the communications among VSN nodes frequently rely on other nodes. Functionality expected of a node will be dependent on whether or not it is currently a member of a VSN. Those that are not in the VSN need to support the maintenance of VSNs, while those within the VSN need to carry out tasks such as profile detection, pattern recognition, and tracking.

The VSN maintenance functions include:

- adding and deleting nodes (decision made by nodes other than that being added/deleted),

- nodes entering and leaving VSN (decision made by node itself),

- broadcast within VSN,

- join two VSNs (ex. when two plumes merge),

- splitting VSNs (ex. plume broken into parts), and

- deriving contours of boundaries.

The supporting nodes (i.e., nodes that do not belong to the VSN at present) need to provide efficient message exchanges among the sensor nodes for implementation of those functions. These functions have to be implemented with minimal overhead, while taking other limiting factors in wireless sensor networks into account.

In addition, efficiently managing the membership changes in the VSN is critical for energy conservation. Sensors have different role assignments in the context of plume monitoring. For instance, sensors within the VSN actively participate in sensing for profile detection, pattern recognition and tracking; sensors outside the VSN may help relay data from VSN members to the server, or simply remain asleep, depending on whether or not they are on the path to the server. Different roles may impose different burdens on nodes. For example, nodes within the VSN consume more energy due to sensing and communicating its own readings to the server. These different sensor roles must be taken into account in supporting real-time plume monitoring, since nodes with certain critical roles may affect application level quality to a high degree while overburdened nodes might be more liable to energy starvation. Moreover, these roles need not be statically assigned to nodes in the system, since sensor roles will be changed when the VSN membership changes. Existing research has shown that optimal sensor network lifetime can be achieved by assuming an optimal sequence of feasible role assignments for each node, where the roles are sensing, relay and aggregator . Instead of maximizing sensor network lifetime, we aim to achieve the best monitoring quality while minimizing energy consumption. We plan to use plume monitoring as a driving application to determine optimal sensor role assignment strategy.

This strategy can be realized by using the generic framework for assigning roles in sensor networks . Furthermore, the application-aware role assignments provide a guideline for sensor state transitions (i.e., switching among different power saving states), which can further be used to drive MAC layer protocols. We expect that the integrated application driven network self-organization, sensor role assignments and sensor state management will ultimately meet the requirements from plume monitoring while minimizing sensor energy consumption.

To conclude the concept of virtual sensor networks over wireless sensor networks will result in sensor networking protocols and data processing algorithms that would have wide applicability, especially in environments where multiple geographically overlapping sensor networks are deployed. In this case, the concept of VSN allows the different sensor networks in the area to operate as VSNs, but make use of nodes (e.g., those that are not power limited) from other sensor networks in the neighborhood. VSN support will simplify application deployment, enhance performance and scalability, facilitate resource sharing, and provide a degree of physical topology independence in wireless sensor networks. The protocols and algorithms for VSN will provide for formation of VSNs, communication functions such as broadcast and anycast among nodes, and adding and dropping nodes from VSN. Efficient implementation of this functionality is key to the success of the proposed approach. New application deployment will be simplified, for example, due to the fact that nodes implementing new application-specific processing can be intermixed with existing nodes. Performance and scalability enhancements can be expected due to the fact that the distributed data processing algorithms can run in the appropriate subset of nodes. We anticipate VSNs to play a key role in emerging sensor based infrastructure.

Location Estimation in Sensor Networks

Dramatic advances in radio frequency (RF) and micro-electro-mechanical systems (MEMS) IC design have made possible the use of large networks of wireless sensors for a variety of new monitoring and control applications. For example, smart structures will actively respond to earthquakes to make buildings and bridges safer, and constantly monitor for cracks or structural problems . Precision agriculture will reduce costs and environmental impact by watering and fertilizing exactly where necessary, and will improve quality by monitoring storage conditions after harvesting [102, 39]. Condition-based maintenance will direct equipment servicing exactly when and where needed based on data from wireless sensors. Traffic monitoring systems will better control stoplights and inform motorists of alternate routes in case of traffic jams. Environmental monitoring networks will sense air, water and soil quality and identify the source of pollutants in real time. A wide variety of such applications have been enabled by the promise of inexpensive networks of wireless sensors.

Automatic estimation of physical location of the sensors in these wireless networks is a key enabling technology. The overwhelming reason is that a sensor's location must be known for its data to be meaningful. If a system is set up to respond locally to changes in sensor data, then it must know where those changes are occurring. In many cases, location itself is the data that needs to be sensed - localization can be the driving need for wireless sensor networks in applications such as warehousing and manufacturing logistics, in which radio tagged parts and equipment must be able to be accurately located at all times. Also, sensor location information, if it is accurate enough, can be extremely useful for scalable, 'geographic' routing algorithms.

Motivation

Many civilian and military applications require monitoring that can identify objects in a specific area, such as monitoring the front entrance of a private house by a single camera. Monitored areas that are large relative to objects of interest often require multiple sensors (e.g., infra-red detectors) at multiple locations. A centralized observer or computer application monitors the sensors. The communication to power and bandwidth requirements call for efficient design of the sensor, transmission, and processing.

The *CodeBlue system* of Harvard university is an example where a vast number of sensors distributed among hospital facilities allow staff to locate a patient in distress. In addition, the sensor array enables online recording of medical information while allowing the patient to move around. Military applications (e.g. locating an intruder into a secured area) are also good candidates for setting a wireless sensor network.

Setting

Let θ denote the position of interest. A set of N sensors acquire measurements $x_n = \theta + w_n$ contaminated by an additive noise w_n owing some known or unknown probability density function (PDF). The sensors transmit measurements to a central processor. The n th sensor encodes x_n by a function $m_n(x_n)$. The application processing the data applies a pre-defined estimation rule $\hat{\theta} = f(m_1(x_1), \cdot, m_N(x_N))$. The set of message functions $m_n, 1 \leq n \leq N$ and the fusion rule $f(m_1(x_1), \cdot, m_N(x_N))$ are designed to minimize estimation error. For example: minimizing the mean squared error (MSE), $\mathbb{E} \| \grave{e} - \hat{e} \|^2$.

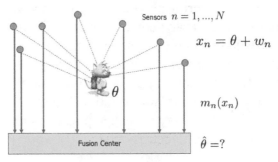

Sensors $n = 1, ..., N$

$x_n = \theta + w_n$

$m_n(x_n)$

$\hat{\theta} = ?$

Fusion Center

Ideally, sensors transmit their measurements x_n right to the processing center, that is $m_n(x_n) = x_n$. In this settings, the maximum likelihood estimator (MLE) $\hat{\theta} = \dfrac{1}{N}\sum_{n=1}^{N} x_n$ is an unbiased estimator whose MSE is $\mathbb{E}\|\theta - \hat{\theta}\|^2 = var(\hat{\theta}) = \dfrac{\sigma^2}{N}$ assuming a white Gaussian noise. $w_n \sim \mathcal{N}(0, \sigma^2)$. The next sections suggest alternative designs when the sensors are bandwidth constrained to 1 bit transmission, that is $m_n(x_n) = 0$ or 1.

Known Noise PDF

We begin with an example of a Gaussian noise $w_n \sim \mathcal{N}(0, \sigma^2)$, in which a suggestion for a system design is as follows

$$m_n(x_n) = I(x_n - \tau) = \begin{cases} 1 & x_n > \tau \\ 0 & x_n \leq \tau \end{cases}$$

$$\hat{\theta} = \tau - F^{-1}\left(\frac{1}{N}\sum_{n=1}^{N} m_n(x_n)\right), \quad F(x) = \frac{1}{\sqrt{2\pi}\sigma}\int_{x}^{\infty} e^{-w^2/2\sigma^2}\, dw$$

Here τ is a parameter leveraging our prior knowledge of the approximate location of θ. In this design, the random value of $m_n(x_n)$ is distributed Bernoulli~ $(q = F(\tau - \theta))$. The processing center averages the received bits to form an estimate $\hat{q}$ of q, which is then used to find an estimate of θ. It can be verified that for the optimal (and infeasible) choice of $\tau = \theta$ the variance of this estimator is $\dfrac{\pi\sigma^2}{4}$ which is only $\pi/2$ times the variance of MLE without bandwidth constraint. The variance increases as τ deviates from the real value of θ, but it can be shown that as long as $|\tau - \theta| \sim \sigma$ the factor in the MSE remains approximately 2. Choosing a suitable value for τ is a major disadvantage of this method since our model does not assume prior knowledge about the approximated location of θ. A coarse estimation can be used to overcome this limitation. However, it requires additional hardware in each of the sensors.

A system design with arbitrary (but known) noise PDF can be found in. In this setting it is assumed that both θ and the noise w_n are confined to some known interval $[-U, U]$. The estimator of also reaches an MSE which is a constant factor times $\dfrac{\sigma^2}{N}$. In this method, the prior knowledge of U replaces the parameter τ of the previous approach.

Unknown Noise Parameters

A noise model may be sometimes available while the exact PDF parameters are unknown (e.g. a Gaussian PDF with unknown σ). The idea proposed in for this setting is to use two thresholds τ_1, τ_2, such that $N/2$ sensors are designed with $m_A(x) = I(x - \tau_1)$, and the other $N/2$ sensors use $m_B(x) = I(x - \tau_2)$. The processing center estimation rule is generated as follows:

$$\hat{q}_1 = \frac{2}{N}\sum_{n=1}^{N/2} m_A(x_n), \quad \hat{q}_2 = \frac{2}{N}\sum_{n=1+N/2}^{N} m_B(x_n)$$

$$\hat{\theta} = \frac{F^{-1}(\hat{q}_2)\tau_1 - F^{-1}(\hat{q}_1)\tau_2}{F^{-1}(\hat{q}_2) - F^{-1}(\hat{q}_1)}, \quad F(x) = \frac{1}{\sqrt{2\pi}}\int_x^{\infty} e^{-v^2/2}\,dw$$

As before, prior knowledge is necessary to set values for τ_1, τ_2 to have an MSE with a reasonable factor of the unconstrained MLE variance.

Unknown noise PDF

We now describe the system design of for the case that the structure of the noise PDF is unknown. The following model is considered for this scenario:

$$x_n = \theta + w_n, \quad n = 1,\ldots,N$$

$$\theta \in [-U,U]$$

$w_n \in \mathcal{P}$, *that is* : w_n *is bounded to* $[-U,U], \mathbb{E}(w_n) = 0$

In addition, the message functions are limited to have the form

$$m_n(x_n) = \begin{cases} 1 & x \in S_n \\ 0 & x \notin S_n \end{cases}$$

where each S_n is a subset of $[-2U,2U]$. The fusion estimator is also restricted to be linear, i.e. $\hat{\theta} = \sum_{n=1}^{N} \alpha_n m_n(x_n)$.

The design should set the decision intervals S_n and the coefficients α_n. Intuitively, we would allocate $N/2$ sensors to encode the first bit of θ by setting their decision interval to be $[0,2U]$, then $N/4$ sensors would encode the second bit by setting their decision interval to $[-U,0]\cup[U,2U]$ and so on. It can be shown that these decision intervals and the corresponding set of coefficients α_n produce a universal δ-unbiased estimator, which is an estimator satisfying $|\mathbb{E}(\theta - \hat{\theta})| < \delta$ for every possible value of $\theta \in [-U,U]$ and for every realization of $w_n \in \mathcal{P}$. In fact, this intuitive design of the decision intervals is also optimal in the following sense. The above design requires $N \geq \lceil \log\frac{8U}{\delta} \rceil$ to satisfy the universal δ-unbiased property while theoretical arguments show that an optimal (and a more complex) design of the decision intervals would require $N \geq \lceil \log\frac{2U}{\delta} \rceil$, that is: the number of sensors is nearly optimal. It is also argued in that if the targeted MSE $\mathbb{E}\|\theta - \hat{\theta}\| \leq \epsilon^2$ uses a small enough ϵ then this design requires a factor of 4 in the number of sensors to achieve the same variance of the MLE in the unconstrained bandwidth settings.

Additional Information

The design of the sensor array requires optimizing the power allocation as well as minimizing the communication traffic of the entire system. The design suggested in incorporates probabilistic quantization in sensors and a simple optimization program that is solved in the fusion center only once. The fusion center then broadcasts a set of parameters to the sensors that allows them to finalize their design of messaging functions $m_n(\cdot)$ as to meet the energy constraints. Another work employs a similar approach to address distributed detection in wireless sensor arrays.

Key Distribution in Wireless Sensor Networks

Wireless sensor networks (or simply sensor networks throughout this paper) have been proven to be useful in a large number of applications. However, the nature of limited resources on sensor nodes restricts the use of conventional security techniques in sensor networks. The most concerned security issues in sensor networks are the same with the traditional Internet. For example, confidentiality, authenticity, and integrity are also important in sensor networks, but due to the feature of sensor networks, some particular attacks have been carried out. For instance, Wormhole attack, Node replication attack, and Sybil attack do not exist in the traditional networks. In this paper, we will focus on the confidentiality.

One of major concerns in the sensor network applications is how the confidentiality of the sensed data and the control message exchanged among sensor nodes can be guaranteed [1–4]. From conventional wisdom, we know that the use of encryption functions can offer the confidentiality. Here, a critical issue of how to properly distribute the secret keys among sensor nodes emerges. Note that although asymmetric key cryptography has been proven to be usable in sensor networks, as the time required by the operations in asymmetric key cryptography is still the order of seconds, symmetric key cryptography is still preferable despite the key management problem it incurs.

Evaluation Metrics

To evaluate the efficiency and effectiveness of key distribution schemes in sensor networks, the following metrics, are considered usually

- Resilience against sensor compromises—Because sensor nodes are assumed to be not tamper resistant, the adversary may physically compromise the sensor nodes, attempting to retrieve the credentials stored in the compromised sensor nodes and then launch further insider attacks. In the context of key distribution, the adversary may try to infer the key shared between the uncompromised sensor nodes. Thus, the resilience against sensor compromises defined to evaluate the ability to preserve the confidentiality of the data generated and exchanged between the uncompromised sensor nodes.

- Resource efficiency—Because sensor nodes are resource constrained, a good key distribution scheme should not consume a large amount of resources. The resources here could be computation power, communication capability, and storage space. For example, the asymmetric key cryptography is not considered because of its heavy demand of computation requirement. Because communication dominates the energy consumption of sensor nodes, the number of exchanged messages required to establish the secret key should be as small as possible. The memory in sensor nodes is only of tens of kilobytes, implying that the key distribution scheme cannot store too many keying materials in sensor nodes.

- Connectivity—The connectivity here is denoted as the probability that a pair of sensor nodes can find their shared key. As it will be shown later, a large number of key distribution schemes exploit the technique of probabilistic key sharing to enable sensor nodes to compute their shared key. However, it also implies that not every pair of sensor nodes can have a shared key. Obviously, the connectivity should be as large as possible.

- Adaptability—Numerous key distribution schemes have been proposed. Different key distribution schemes have different assumptions. For example, some may assume the use of mobile robots to enhance the connectivity, and some may assume the use of secure bootstrapping time to strengthen the resilience against the sensor compromises. On the other hand, network settings can also be different. For example, some networks consist of mobile sensor nodes, or some sensor nodes can be aware of their locations. The adaptability here is to evaluate whether the key distribution schemes can be adaptable to the different network settings.

Key Distribution Schemes

Key predistribution is the method of distribution of keys onto nodes before deployment. Therefore, the nodes build up the network using their secret keys after deployment, that is, when they reach their target position.

Key predistribution schemes are various methods that have been developed by academicians for a better maintenance of PEA management in WSNs. Basically a key predistribution scheme has 3 phases:

1. Key distribution

2. Shared key discovery

3. Path-key establishment

During these phases, secret keys are generated, placed in sensor nodes, and each sensor node searches the area in its communication range to find another node to communicate.

A secure link is established when two nodes discover one or more common keys (this differs in each scheme), and communication is done on that link between those two nodes. Afterwards, paths are established connecting these links, to create a connected graph. The result is a wireless communication network functioning in its own way, according to the key predistribution scheme used in creation.

There are a number of aspects of WSNs on which key predistribution schemes are competing to achieve a better result. The most critical ones are: local and global connectivity, and resiliency.

Local connectivity means the probability that any two sensor nodes have a common key with which they can establish a secure link to communicate.

Global connectivity is the fraction of nodes that are in the largest connected graph over the number of all nodes.

Resiliency is the number of links that cannot be compromised when a number of nodes(therefore keys in them) are compromised. So it is basically the quality of resistance against the attempts to hack the network. Apart from these, two other critical issues in WSN design are computational cost and hardware cost. Computational cost is the amount of computation done during these phases. Hardware cost is generally the cost of the memory and battery in each node.

Keys may be generated randomly and then the nodes determine mutual connectivity. A structured approach based on matrices that establishes keys in a pair-wise fashion is due to Rolf Blom. Many variations to Blom›s scheme exist. Thus the scheme of Du et al. combines Blom's key pre-distribution scheme with the random key pre-distribution method with it, providing better resiliency.

Polynomial based Key Pre-distribution

We discuss how to pre distribute a pairwise keys.

The (key) setup server randomly generates a bivariate t-degree polynomial $f(x, y) = \sum_{i,j=0}^{t} a_{ij} x^i y^j$ over a finite field Fq , where q is a prime number that is large enough to accommodate a cryptographic key, such that it has the property of $f(x, y) = f(y, x)$. (We assume all the bivariate polynomials have this property without explicit statement.) Each sensor is associated to a unique ID in this system. For each sensor 'i', the setup server computes a polynomial share of f(x, y), that is, f(i, y). For any two sensor nodes i and j, node i can compute the common key f(i, j) by evaluating f (i, y) at point j, and node j can compute the same key f (j, i) = f (i, j) by evaluating f(j, y) at point i.

Polynomial Pool based Key Pre-distribution

As the name suggest, in this technique we have a pool of randomly generated bivariate polynomial; the method used for generating this bivariate polynomial is based on the polynomial based key pre-distribution as discussed above. There are two cases of the polynomial pool. When the polynomial pool has only one polynomial, the general framework degenerates into the polynomial-based key pre distribution. When all the polynomials are 0-degree ones, the polynomial pool degenerates into a key pool Random Subset assignment key pre-distribution.

Grid based Key Pre-distribution

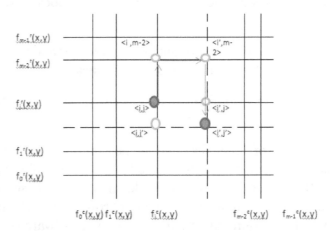

Assume a sensor network with at most N sensor nodes. The grid based key pre distribution scheme then constructs a m×m grid with a set of 2m polynomials $\{ f_{i^c}(x,y), f_{j^r}(x,y)\}\ i=0,...,m-1\}$, where m = $\lceil\sqrt{N}\rceil$. Each row j in the grid is associated with a polynomial $f_{i^c}(x,y)$, and each column i is associated with a polynomial $f_{j^r}(x,y)$. The setup server assigns each sensor in the network to a unique intersection in this grid. For the sensor at the coordinate (i, j), the setup server distributes the polynomial shares of $f_{i^c}(x,y)$, and $f_{j^r}(x,y)$ to the sensor. As a result, sensor nodes can perform share discovery and path discovery based on this information and eventually can pre-distribute the keys.

Hyper-cube Multivariate Scheme (HMS)

It is basically a Threshold based scheme, a number of multivariate polynomial is assigned to every point on the hypercube, and a hypercube is designed in the multidimensional space. The points on the hypercube are uniquely assigned to the sensors. Using this technique, a direct key is established between any two sensors at Hamming distance of one from each other. Other sensors are also able to establish indirect keys.

Key Pre-distribution Scheme using Combinatorial Design

This scheme is a deterministic key pre-distribution in WSN. First to understand the design we need to define some terms

Definition 1: Balanced Incomplete Block Design (BIBD): A set system or design is a pair (X,A), where A is a set of subsets of X, called blocks. The elements of X are called varieties. A BIBD (v,b,r,k,λ) is a design which satisfy the following conditions:

i. $|X| = v$. $|A| = b$,

ii. Each subset in A contains exactly k elements

iii. Each variety in X occurs in r many blocks

iv. Each pair of varities in X is contained in exactly λ blocks in A.

A BIBD(v,b,r,k,λ) design can be represented by a incidence matrix M = [mij] of dimension v x b with entries 0 and 1, mij = 1, if the ith variety is present in the jth block and 0 otherwise. Definition 2 : suppose that (X,A) is a system, where

$$X = \{xi : 1 \leq i \leq v\}$$

And

$$A = \{Aj : 1 \leq j \leq b\}$$

The dual set system of (X,A) is any set isomorphic to the set system (X',A') where

$$X' = \{x'j : 1 \leq j \leq b\}.$$

$$A' = \{A'i : 1 \leq i \leq v\}.$$

And where

$$X'_j \in A'_i \quad \Leftrightarrow \quad x_i \in A_j$$

It follows that if we take the dual of a BIBD(v,b,r,k,λ), we arrive at a design containing b varities, v blocks each block containing exactly r varieties and each variety occurring in exactly k blocks. We also note that any two blocks contain λ elements in common.

Definition 3 Symmetric BIBD or SBIBD: When v = b , the BIBD is called SBIBD and denoted by SB[v, k;λ].

Definition 4 Pairwise balance design (PBD): It is a design in which each pair of points occurs in λ blocks, for some constant λ, called the index of the design.

Definition 5: The intersection number between any two blocks is the number of elements common to the blocks.

Definition 6 : Let the intersection numbers between any of the blocks in a BIBD be $\mu 1$, $\mu 2$,.... μx. let M = { μi : i = 1,2,....,x } let μ = max { μ_1, μ_2,.... μ_x } . μ is called the linkage of the design. Note: For a SBIBD, $|M| = 1$ and, $\mu = \lambda$.

Design: A BIBD (v,b,r,k,λ) can be mapped to a sensor network containing v keys in the keypool. There are b sensor nodes, each node containing k keys are each key occurring in r nodes. Any pair of keys occurs in λ blocks. Consider two BIBDs : $D_1 = \left(v_1, b_1, r_1, k_1, \lambda_1\right)$ and D2 $=\left(v_2, b_2, r_2, k_2, \lambda_2\right)$. Let M1 $= \left[m_{ij}'\right]$ and M2 $= \left[m_{ij}''\right]$ be the respective incidence matrices,b therefore the dimensions of M_1 and M_2 are $v_1 \times b_1$ and $v_2 \times b_2$ respectively. A requirement of the design is that $k_1 = v_2$. The matrix M is constructed in the following way

1. For every column j of M_1 replace m_{ij}' by a row of M_2 if $m_{ij}' = 1$. For each i replace m_{ij}' by a different row of M_2.

2. For every column j of M_1 replace m_{ij}' by a row vector of length b_2 containing all zero's if $m_{ij}' = 0$.

The result of the following operations is a matrix M of dimension $v_1 \times b_1 b_2$. We call the design D and represent it by $D = D_1 \bowtie D_2$. It is said D is the expanded design of D_1 and D_2. The blocks in D which arise from a given block in D1 are said to belong to the same group. Now we map this design D to a sensor network consisting of $b_1 b_2$ nodes, each node consisting of k_2 keys. The key-pool consists of v_1 keys .

A Robust Key Pre-distribution Protocol for Multi-phase WSN

From the components of sensor node we see that they are battery operated and their life time is much smaller as compared to the lifetime of the whole network and hence we need to deploy new sensor nodes for the disappearing ones in the network to assure good network connectivity. The process of deploying new sensors at regular period is called generations. The time between two successive generations is called generation period. Gw - generation window:- a number which is assumed that a sensor lifetime is bounded of generations. The generation period is assumed to be 1. The time period for a node deployed at generation j to establish a secret channel with any other sensor deployed is $|jk, j+k|$ where k is a integer and its assume that k = Gw in this scheme. If the value of k is less than Gw the newly deployed node can establish a secure channel with a subset of the network sensors. Each sensor is assigned A two different key rings: FKR_A and BKR_A , where a key ring is a subset of key pool namely $FKR_A \subset FKP$ and $BKR_A \subset BKP$. The abbreviation and their full form are listed below

Generation window, n:last generation of the network, A: sensor $kr_A^j = \left(FKR_A^j, BKR_A^j\right)$ key ring of A at gen. j, FKR_A^j : forward key ring of A at gen. j, BKR_A^j : backward key ring of A at gen. j, m:key ring size, FKR^j : Forward key pool at gen. j, BKR^j : Backward key pool at gen. j, P:key pool size, fkj t $\in$ FKR^j : t-th f key at gen j, bkj t $\in$ BKR^j : t-th b key at gen j, h:secure hash function $h : \{0,1\}^* \rightarrow \{0,1\}^{160}$, f:hash function $f : \{0,1\}^* \rightarrow \{0,1\}^{\log_2}(P)$, RKP: key management defined in , Rok: Robust key predistribution scheme.

Key Pool Generation:

In this scheme the key pool evolve with time and are updated at each generation. Initially the forward key pool contains P/2 random keys. Each key is updated by hashing the current key with secure hash function h, such as SHA1. When the network is deployed the forward key pool is defined as

$$FKP^0 = \left\{ fk^0{}_1, fk^0{}_2, \ldots, fk^0 p/2 \right\}$$

$fk^0{}_1$ is randomly generated. At generation j+1, the fkeys are refreshed as follows

$$FKP^{j+1} = \left\{ fk^{j+1}{}_1, fk^{j+1}{}_2, \ldots, fk^{j+1} p/2 \right\}$$

Where $fk^{j+1}t = h\left(fk^j t \right)$.

The backward key pool is generated in the similar fashion

Key ring assignments:

By using a pseudorandom function, subkeys are assigned to each node A deployed at generation j and each node is configured with m/2 subkeys from the forward and backward key pools. The first subkey of the forward key ring will be the subkey with index f(idA||1||j) of the forward key pool, where f(.) is for example a hash function with range[1;m]. The first backward subkey of backward key ring will be the subkey with index f(idA||1||j) of the backward key pool and so on.

We can say that a node A is configured with a key ring, $kr^j{}_A = \left(FKR^j{}_A, BKR^j{}_A \right)$ defined as follows

$$FKR^j A = \left\{ fk^j t \mid t = f(idA||1||j), i = 1,2,\ldots,m/2 \right\}$$

$$BKR^j A = \left\{ bk^{j+Gw-1}t \mid t = f(idA||1||j), i = 1,2,\ldots,m/2 \right\}$$

Where m is the size of the whole key ring.

Depending on the deployment of the owner sensor the key rings are strictly bounded to them, thus the nodes need to be loosely time synchronized. Updation of a key ring by a node A for the generations of i can only be between j and j+Gw. The lifetime of key ring is limited. Node A cannot compute the bkeys for the generation after generation j+Gw, it cannot update krA beyond generation j+Gw, it also cannot recover any fkeys for the generation before j and it cannot compute krA for the generation before j. This factor limits the power of the attacker and attacker can use the keys for a limited time span.

Establishing a secure channel:

Neighbor discovery process starts when a node is deployed, consider a node A deployed at generation j, it broadcast a message containing its identifier idA and its generation, the nodes receiving its message are able to reconstruct the list of the key indexes in krA and identify the keys that they have in common. Consider a neighbor node B which constructs the set $\{ t \mid t = f(idA\|l\|j) , l = 1, 2, ..., m/2\}$ of the key indexes and scan to see if it shares at least one index t with A. If it shares, then node B replies with its identifier idB and its generation. Thus A identifies the list t1, t2,...., of all the key indexes in common. After this step both the nodes calculate their overlapping generations which is the set of generations in which they can communicate. The overlapping generation will be all the generation in the interval |j,i+Gw| where i is the generation when B was deployed and j when A was deployed and i ≤ j. With the common key indexes (t1,t2,...,tz)between the two nodes they can communicate their forward and backward keys and eventually compute their secret key which is

$$K_{AB} = h\left(fk^j\, t1 \,\|\, bk^{i+Gw-1}t1 \,\|\, fk^j\, t2 \,\|\, bk^{j+Gw-1}t2 \,\|\, ... \,\|\, fk^j\, tz \,\|\, bk^{i+Gw-1}tz \right)$$

The key kAB can be used to establish a secure channel between A and B. The forward key should be erased by a node at each new generation, this prevents adversary from compromising forward keys.

Matrix based Key Agreement Scheme

The basic algorithm is as follows :

Pre-deployment:

A symmetric random matrix K with elements in Zp is to be chosen. We need to find two matrix X and Y such that XY=K Assign nodes with random row - column pairs from X and Y, for example if a node i receive the rth row of X then it also receives the rth column of Y.

Key Agreement:

Any two nodes i and j agree on an encryption key by exchanging their columns of Y and compute key $K_{ij} = \text{row}_{node(i)}(Y) = \text{row}_{node(j)}(X).\text{col}_{node(i)}(Y) = K_{ji}$, $K_{ij} = K_{ji}$ since K is symmetric. $\text{row}_{node(i)}(X)$ refer to row of X that was assigned to node i .

Commuting matrices can also be used; it eliminates the requirement of K being symmetric. The algorithm is as follows

Pre-deployment:

Choose two q x q matrices X and Y such that XY = YX and Y is symmetric.

Randomly pick r from a uniform distribution over [1,q].

Assign node i randomly chosen rth row and column of X and the rth column of Y

Two nodes I and j agree on key as follows

Node i send its column of Y to node j.

Node j send its column of Y to node i

Node i computes $Kij = row_{node(i)}(X).col_{node(j)}(Y)$ and node j computes $Kij = col'_{node(i)}(Y).col_{node(j)}(X)$

Node i computes $Kji = col'_{node(j)}(Y).col_{node(i)}(X)$ .and node j computes $Kji = row_{node(j)}(X).col_{node(i)}(Y)$

Key used is computed as K=Hash(Kij|| Kji).

Where $col'_{node(i)}(Y)$ is the transpose of column Y assigned to node i.

Scheme 1

Let K be a matrix of size q x q , therefore X and Y are of size q x m and m x q respectively. Number of elements in the upper triangle of matrix including the diagonal is q(q+1)/2 generated from 2qm elements. For each key there are p possibilities which are equally likely.

Scheme 2

In this scheme the commuting matrices X and Y requires to be a square matrix q x q where q ≤ N where N is the number of nodes in a network. K = Hash (Kij|| Kji) is the common key between nodes i and j.

Example: We will show a simple example of Scheme 1 using symmetric matrix. Let K be a symmetric, $K = \begin{bmatrix} 2 & 4 \\ 4 & 5 \end{bmatrix}$ then we find 2 matrices X and Y such that $XY = K$ $X = \begin{bmatrix} 2 & 4 \\ 4 & 5 \end{bmatrix}$ $Y = \begin{bmatrix} 1 & 2 \\ 1 & 1 \end{bmatrix}$, then we can calculate for i = 1 and j = 2, $Kij = row_1(X).col_2(Y) = 4$ and $Kji = row_2(X).col_1(Y) = 4$.

The Piggy Bank based Methods

The idea of locking and sealing a piggy bank to transport message is extended to data communication . Once a secret message is inserted into the piggy bank and locked, than the message is not accessible until the box is opened with an appropriate key. We first discuss a three-stage protocol in which both parties use lock protected by tamper-proof seal of each parties and then discuss the piggy bank trope. Suppose Alice wants to send a message to Bob, Alice inserts his message in a locked box which is transported to Bob; Bob in turn puts his own lock on the box which is transported back to Alice. Alice unlocks the lock using his key Ka and sends the box again to Bob and then Bob unlocks it with his key Kb. By unlocking with their key they ensure that the lock is not tampered.

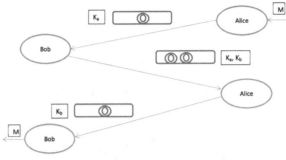

Three-stage protocol using locks

The Piggy Bank Scheme: The steps involved in this trope are as shown in figure below

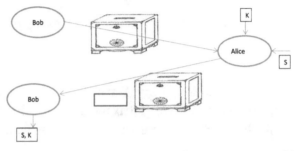

Piggy Bank Cryptographic Trope; the secret letter is represented by ▭

An empty locked piggy bank is sent by Bob to Alice; Alice on receiving the box deposits the secret message along with the decryption key of a coded letter and sends it back to Bob. Alice also sends an encrypted letter to authenticate the contents of the locked piggy bank box. Bob opens the box, obtains the secret and also reads the coded letter.

Notice that the communication from Alice to Bob consists of two separate messages. One of these, represented by the box can be pre distributed within the network. This reduces the computation burden on the second transformation which now can be used with reduced complexity in a manner that is appropriate for the reduced computational capacity of the sensor nodes.

In order to understand how it can be implemented in real system let us explain the working with the help of an example. The protocol is implemented in 3 steps:

Step 1: Bob selects a random number say 11 and the piggy bank transformation is represented by a one-way transformation $f(11) = 11^{1037} \mod 323 = 7$ where 323 is a composite number with factors known only to Bob; 1037 is the publicly known encryption exponent. The composite number required for computation is provided by Bob to Alice by some secured channel.

Step 2: Bob sends f(11) = 7 to Alice who multiples it with her secret key 3 and sends f(11)*3 + 17 mod 323 = 38 in one communication and $f(3) = 3^{1037} \mod 323 = 29$ in another communication.

Step 3: Bob uses his secret inverse transformation function to recover the secret key of Alice; first he recovers 3 and then 17. The inverse transformation is as follows 29^5 mod 323 = 3

3 * 7 + K = 38 therefore K = 17

Thus we can see from the above calculation Bob retrieves the secret keys of Alice. The above steps can be shown pictorially as follows

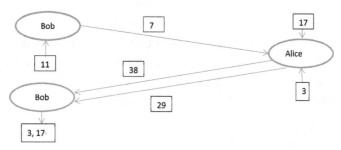

Piggy Bank Cryptographic Trope implementation

Since a sensor node has limited resources, normally public cryptography is not feasible. However, we can use the piggy bank approach in WSN by reducing the computation involved in the process. This can be done by hardwiring one of the secret keys. Once a secret key is hardwired, we can use smaller composite numbers without weakening the level of security.

References

- Dargie, W. and Poellabauer, C. (2010). Fundamentals of wireless sensor networks: theory and practice. John Wiley and Sons. pp. 168–183, 191–192. ISBN 978-0-470-99765-9

- Peiris, V. (2013). "Highly integrated wireless sensing for body area network applications". SPIE Newsroom. doi:10.1117/2.1201312.005120

- "Wireless temperature sensor for Data Centers". ServersCheck. Archived from the original on 2016-10-11. Retrieved 2016-10-09

- Niazi, Muaz; Hussain, Amir (2011). "A Novel Agent-Based Simulation Framework for Sensing in Complex Adaptive Environments" (PDF). IEEE Sensors Journal. 11 (2): 404–412. doi:10.1109/jsen.2010.2068044. Archived from the original(PDF) on 2011-07-25

- Sohraby, K., Minoli, D., Znati, T. (2007). Wireless sensor networks: technology, protocols, and applications. John Wiley and Sons. pp. 203–209. ISBN 978-0-471-74300-2

- Tony O'Donovan; John O'Donoghue; Cormac Sreenan; David Sammon; Philip O'Reilly; Kieran A. O'Connor (2009). A Context Aware Wireless Body Area Network (BAN) (PDF). Pervasive Computing Technologies for Healthcare, 2009. doi:10.4108/ICST.PERVASIVEHEALTH2009.5987. Archived(PDF) from the original on 2016-10-09

- K. Saleem; N. Fisal & J. Al-Muhtadi (2014). "Empirical studies of bio-inspired self-organized secure autonomousRouting protocol". Sensors Journal IEEE. 14: 1–8. doi:10.1109/JSEN.2014.2308725. Archived from the original on 2018-05-03

4

Wireless Communications: Technologies and Devices

Wireless communication refers to the process of information exchange between two or more remote points that are not connected by an electrical conductor. This transfer of information is achieved by radio waves. All the diverse technologies and devices of wireless communications have been carefully analyzed in this chapter such as mobile telephony, Bluetooth, Wi-Fi and infrared communication.

Mobile Telephony

A mobile phone is an electronic device used for mobile telecommunications over a cellular network of specialized base stations known as cell sites. A cell phone offers full Duplex Communication and transfer the link when the user moves from one cell to another. As the phone user moves from one cell area to another, the system automatically commands the mobile phone and a cell site with a stronger signal, to switch on to a new frequency in order to keep the link.

Mobile phone is primarily designed for Voice communication. In addition to the standard voice function, new generation mobile phones support many additional services, and accessories, such as SMS for text messaging, email, packet switching for access to the Internet, gaming, Bluetooth, camera with video recorder and MMS for sending and receiving photos and video, MP3 player, radio and GPS.

The different types of mobile communication systems are mobile two-way radio, public land radio, mobile telephone and amateur (HAM) radio.

Mobile two-way radios are one-to-many communication systems that operate in half-duplex mode, i.e., push to talk. The most common among this type is citizen band (CB) radio, which uses amplitude modulation (AM). It operates in the frequency range of 26-27.1 MHz having 40 channels of 10 kHz. It is a non-commercial service that uses a press-to-talk switch. It can be amplitude modulated having double-sideband suppressed carrier or single-sideband suppressed carrier.

Public land mobile radio is a twoway FM radio system, used in police, fireand municipal agencies. It is limited to small geographical areas.

Amateur (HAM) radios cover a broad frequency band from 1.8 MHz to above 30 MHz. These include continuous wave (CW), AM, FM, radio teleprinter, HF slow-scan still picture TV, VHF or UHF slow-scan or fast-scan TV, facsimile, frequency-shift keying and amplitude-shift keying.

Mobile telephones offer full-duplex transmission. These are one-to-one systems that permit two simultaneous transmissions. For privacy, each mobile unit carries a unique telephone number.

Amateur (HAM) radios cover a broad frequency band from 1.8 MHz to above 30 MHz. These include continuous wave (CW), AM, FM, radio teleprinter, HF slow-scan still picture TV, VHF or UHF slow-scan or fast-scan TV, facsimile, frequency-shift keying and amplitude-shift keying.

Signal Frequency in Cell Phone

The cellular system is the division of an area into small cells. This allows extensive frequency reuse across that area, so that many people can use cell phones simultaneously. Cellular networks has a number of advantages like increased capacity, reduced power usage, larger coverage area, reduced interference from other signals etc.

FDMA and CDMA Systems

Frequency Division Multiple Access (FDMA) and Code Division Multiple Access (CDMA) were developed to distinguish signals from several different transmitters. In FDMA, the transmitting and receiving frequencies used in each cell are different from the frequencies used in the neighboring cells. The principle of CDMA is more complex and the distributed transceivers can select one cell and listen to it. Other methods include Polarization Division Multiple Access (PDMA) and Time Division Multiple Access (TDMA). Time division multiple access is used in combination with either FDMA or CDMA to give multiple channels within the coverage area of a single cell.

Cellular Systems

Mobile phones receive and send radio signals with any number of cell site base stations fitted with microwave antennas. These sites are usually mounted on a tower, pole or building, located throughout populated areas, then connected to a cabled communication network and switching system. The phones have a low-power transceiver that transmits voice and data to the nearest cell sites, normally not more than 8 to 13 km (approximately 5 to 8 miles) away. In areas of low coverage, a cellular repeater may be used, which uses a long distance high-gain dish antenna or yagi antenna to communicate with a cell tower far outside of normal range, and a repeater to rebroadcast on a small short-range local antenna that allows any cellphone within a few meters to function properly.

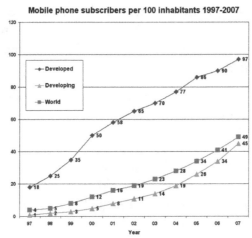

Mobile phone subscriptions, not subscribers, per 100 inhabitants

When the mobile phone or data device is turned on, it registers with the mobile telephone exchange, or switch, with its unique identifiers, and can then be alerted by the mobile switch when there is an incoming telephone call. The handset constantly listens for the strongest signal being received from the surrounding base stations, and is able to switch seamlessly between sites. As the user moves around the network, the "hand-offs" are performed to allow the device to switch sites without interrupting the call.

Cell sites have relatively low-power (often only one or two watts) radio transmitters which broadcast their presence and relay communications between the mobile handsets and the switch. The switch in turn connects the call to another subscriber of the same wireless service provider or to the public telephone network, which includes the networks of other wireless carriers. Many of these sites are camouflaged to blend with existing environments, particularly in scenic areas.

The dialogue between the handset and the cell site is a stream of digital data that includes digitised audio (except for the first generation analog networks). The technology that achieves this depends on the system which the mobile phone operator has adopted. The technologies are grouped by generation. The first-generation systems started in 1979 with Japan, are all analog and include AMPS and NMT. Second-generation systems, started in 1991 in Finland, are all digital and include GSM, CDMA and TDMA.

The nature of cellular technology renders many phones vulnerable to 'cloning': anytime a cell phone moves out of coverage (for example, in a road tunnel), when the signal is re-established, the phone sends out a 're-connect' signal to the nearest cell-tower, identifying itself and signalling that it is again ready to transmit. With the proper equipment, it's possible to intercept the re-connect signal and encode the data it contains into a 'blank' phone—in all respects, the 'blank' is then an exact duplicate of the real phone and any calls made on the 'clone' will be charged to the original account. This problem was widespread with the first generation analogue technology, however the modern digital standards such as GSM greatly improve security and make cloning hard to achieve.

In an effort to limit the potential harm from having a transmitter close to the user's body, the first fixed/mobile cellular phones that had a separate transmitter, vehicle-mounted antenna, and handset (known as *car phones* and *bag phones*) were limited to a maximum 3 watts Effective Radiated Power. Modern *handheld* cellphones which must have the transmission antenna held inches from the user's skull are limited to a maximum transmission power of 0.6 watts ERP. Regardless of the potential biological effects, the reduced transmission range of modern handheld phones limits their usefulness in rural locations as compared to car/bag phones, and handhelds require that cell towers are spaced much closer together to compensate for their lack of transmission power.

Usage

By Civilians

This Railfone found on some Amtrak trains in North America uses cellular technology

An increasing number of countries, particularly in Europe, now have more mobile phones than people. According to the figures from Eurostat, the European Union's in-house statistical office, Luxembourg had the highest mobile phone penetration rate at 158 mobile subscriptions per 100 people, closely followed by Lithuania and Italy. In Hong Kong the penetration rate reached 139.8% of the population in July 2007. Over 50 countries have mobile phone subscription penetration rates higher than that of the population and the Western European average penetration rate was 110% in 2007 (source Informa 2007). Canada currently has the lowest rates of mobile phone penetrations in the industrialised world at 58%.

There are over five hundred million active mobile phone accounts in China, as of 2007, but the total penetration rate there still stands below 50%. The total number of mobile phone subscribers in the world was estimated at 2.14 billion in 2005. The subscriber count reached 2.7 billion by end of 2006 according to Informa, and 3.3 billion by

November, 2007, thus reaching an equivalent of over half the planet's population. Around 80% of the world's population has access to mobile phone coverage, as of 2006. This figure is expected to increase to 90% by the year 2010.

In some developing countries with little "landline" telephone infrastructure, mobile phone use has quadrupled in the last decade. The rise of mobile phone technology in developing countries is often cited as an example of the leapfrog effect. Many remote regions in the third world went from having no telecommunications infrastructure to having satellite based communications systems. At present, Africa has the largest growth rate of cellular subscribers in the world, its markets expanding nearly twice as fast as Asian markets. The availability of prepaid or 'pay-as-you-go' services, where the subscriber is not committed to a long term contract, has helped fuel this growth in Africa as well as in other continents.

On a numerical basis, India is the largest growth market, adding about 6 million mobile phones every month. It currently has a mobile subscriber base of 937.06 million mobile phones.

Traffic

Since the world is operating quickly to 3G and 4G networks, mobile traffic through video is heading high. It is expected that by end of 2018, the global traffic will reach an annual rate of 190 exabytes/year. This is the result of people shifting to smart phones now-a-days. It is predicted by 2018, mobile traffic will reach by 10 billion connections with 94% traffic comes from Smartphones, laptops and tablets. Also 69% of mobile traffic from Videos since we have high definition screens available in smart phones and 176.9 wearable devices to be at use. Apparently, 4G will be dominating the traffic by 51% of total mobile data by 2018.

By Government Agencies

Law Enforcement

Law enforcement have used mobile phone evidence in a number of different ways. Evidence about the physical location of an individual at a given time can be obtained by triangulating the individual's cellphone between several cellphone towers. This triangulation technique can be used to show that an individual's cellphone was at a certain location at a certain time. The concerns over terrorism and terrorist use of technology prompted an inquiry by the British House of Commons Home Affairs Select Committee into the use of evidence from mobile phone devices, prompting leading mobile telephone forensic specialists to identify forensic techniques available in this area. NIST have published guidelines and procedures for the preservation, acquisition, examination, analysis, and reporting of digital information present on mobile phones can be found under the NIST Publication SP800-101.

In the UK in 2000 it was claimed that recordings of mobile phone conversations made on the day of the Omagh bombing were crucial to the police investigation. In particular, calls made on two mobile phones which were tracked from south of the Irish border to Omagh and back on the day of the bombing, were considered of vital importance.

Further example of criminal investigations using mobile phones is the initial location and ultimate identification of the terrorists of the 2004 Madrid train bombings. In the attacks, mobile phones had been used to detonate the bombs. However, one of the bombs failed to detonate, and the SIM card in the corresponding mobile phone gave the first serious lead about the terrorists to investigators. By tracking the whereabouts of the SIM card and correlating other mobile phones that had been registered in those areas, police were able to locate the terrorists.

Disaster Response

The Finnish government decided in 2005 that the fastest way to warn citizens of disasters was the mobile phone network. In Japan, mobile phone companies provide immediate notification of earthquakes and other natural disasters to their customers free of charge. In the event of an emergency, disaster response crews can locate trapped or injured people using the signals from their mobile phones. An interactive menu accessible through the phone's Internet browser notifies the company if the user is safe or in distress. In Finland rescue services suggest hikers carry mobile phones in case of emergency even when deep in the forests beyond cellular coverage, as the radio signal of a cellphone attempting to connect to a base station can be detected by overflying rescue aircraft with special detection gear. Also, users in the United States can sign up through their provider for free text messages when an AMBER Alert goes out for a missing person in their area.

However, most mobile phone networks operate close to capacity during normal times, and spikes in call volumes caused by widespread emergencies often overload the system just when it is needed the most. Examples reported in the media where this has occurred include the September 11, 2001 attacks, the 2003 Northeast blackouts, the 2005 London Tube bombings, Hurricane Katrina, the 2006 Kiholo Bay earthquake, and the 2007 Minnesota bridge collapse.

Under FCC regulations, all mobile telephones must be capable of dialing emergency telephone numbers, regardless of the presence of a SIM card or the payment status of the account.

Impact on Society

Human Health

Since the introduction of mobile phones, concerns (both scientific and public) have been raised about the potential health impacts from regular use. But by 2008, American mobile phones transmitted and received more text messages than phone calls.

Numerous studies have reported no significant relationship between mobile phone use and health, but the effect of mobile phone usage on health continues to be an area of public concern.

For example, at the request of some of their customers, Verizon created usage controls that meter service and can switch phones off, so that children could get some sleep. There have also been attempts to limit use by persons operating moving trains or automobiles, coaches when writing to potential players on their teams, and movie theater audiences. By one measure, nearly 40% of automobile drivers aged 16 to 30 years old text while driving, and by another, 40% of teenagers said they could text blindfolded.

18 studies have been conducted on the link between cell phones and brain cancer; A review of these studies found that cell phone use of 10 years or more "give a consistent pattern of an increased risk for acoustic neuroma and glioma". The tumors are found mostly on the side of the head that the mobile phone is in contact with. In July 2008, Dr. Ronald Herberman, director of the University of Pittsburgh Cancer Institute, warned about the radiation from mobile phones. He stated that there was no definitive proof of the link between mobile phones and brain tumors but there was enough studies that mobile phone usage should be reduced as a precaution. To reduce the amount of radiation being absorbed hands free devices can be used or texting could supplement calls. Calls could also be shortened or limit mobile phone usage in rural areas. Radiation is found to be higher in areas that are located away from mobile phone towers.

According to Reuters, The British Association of Dermatologists is warning of a rash occurring on people's ears or cheeks caused by an allergic reaction from the nickel surface commonly found on mobile devices' exteriors. There is also a theory it could even occur on the fingers if someone spends a lot of time text messaging on metal menu buttons. In 2008, Lionel Bercovitch of Brown University in Providence, Rhode Island, and his colleagues tested 22 popular handsets from eight different manufacturers and found nickel on 10 of the devices.

Human Behaviour

Culture and Customs

Cellular phones allow people to communicate from almost anywhere at their leisure.

Between the 1980s and the 2000s, the mobile phone has gone from being an expensive item used by the business elite to a pervasive, personal communications tool for the general population. In most countries, mobile phones outnumber land-line phones, with fixed landlines numbering 1.3 billion but mobile subscriptions 3.3 billion at the end of 2007.

In many markets from Japan and South Korea, to Europe, to Malaysia, Singapore, Taiwan and Hong Kong, most children age 8-9 have mobile phones and the new accounts are now opened for customers aged 6 and 7. Where mostly parents tend to give hand-me-down used phones to their youngest children, in Japan already new cameraphones are on the market whose target age group is under 10 years of age, introduced by KDDI in February 2007. The USA also lags on this measure, as in the US so far, about half of all children have mobile phones. In many young adults' households it has supplanted the land-line phone. Mobile phone usage is banned in some countries, such as North Korea and restricted in some other countries such as Burma.

Given the high levels of societal mobile phone service penetration, it is a key means for people to communicate with each other. The SMS feature spawned the "texting" sub-culture amongst younger users. In December 1993, the first person-to-person SMS text message was transmitted in Finland. Currently, texting is the most widely used data service; 1.8 billion users generated $80 billion of revenue in 2006 (source ITU). Many phones offer Instant Messenger services for simple, easy texting. Mobile phones have Internet service (e.g. NTT DoCoMo's i-mode), offering text messaging via e-mail in Japan, South Korea, China, and India. Most mobile internet access is much different from computer access, featuring alerts, weather data, e-mail, search engines, instant messages, and game and music downloading; most mobile internet access is hurried and short.

Because mobile phones are often used publicly, social norms have been shown to play a major role in the usage of mobile phones. Furthermore, the mobile phone can be a fashion totem custom-decorated to reflect the owner's personality and may be a part of their self-identity. This aspect of the mobile telephony business is, in itself, an industry, e.g. ringtone sales amounted to $3.5 billion in 2005. Mobile phone use on aircraft is starting to be allowed with several airlines already offering the ability to use phones during flights. Mobile phone use during flights used to be prohibited and many airlines still claim in their in-plane announcements that this prohibition is due to possible interference with aircraft radio communications. Shut-off mobile phones do not interfere with aircraft avionics. The recommendation why phones should not be used during take-off and landing, even on planes that allow calls or messaging, is so that passengers pay attention to the crew for any possible accident situations, as most aircraft accidents happen on take-off and landing.

Etiquette

Mobile phone use can be an important matter of social discourtesy: phones ringing during funerals or weddings; in toilets, cinemas and theatres. Some book shops, libraries, bathrooms, cinemas, doctors' offices and places of worship prohibit their use, so that other patrons will not be disturbed by conversations. Some facilities install signal-jamming equipment to prevent their use, although in many countries, including the US, such equipment is illegal.

Many US cities with subway transit systems underground are studying or have implemented mobile phone reception in their underground tunnels for their riders, and trains, particularly those involving long-distance services, often offer a "quiet carriage" where phone use is prohibited, much like the designated non-smoking carriage of the past. Most schools in the United States and Europe and Canada have prohibited mobile phones in the classroom, or in school in an effort to limit class disruptions.

A working group made up of Finnish telephone companies, public transport operators and communications authorities has launched a campaign to remind mobile phone users of courtesy, especially when using mass transit—what to talk about on the phone, and how to. In particular, the campaign wants to impact loud mobile phone usage as well as calls regarding sensitive matters.

Use by Drivers

The use of mobile phones by people who are driving has become increasingly common, for example as part of their job, as in the case of delivery drivers who are calling a client, or socially as for commuters who are chatting with a friend. While many drivers have embraced the convenience of using their cellphone while driving, some jurisdictions have made the practice against the law, such as Australia, the Canadian provinces of British Columbia, Quebec, Ontario, Nova Scotia, and Newfoundland and Labrador as well as the United Kingdom, consisting of a zero-tolerance system operated in Scotland and a warning system operated in England, Wales, and Northern Ireland. Officials from these jurisdictions argue that using a mobile phone while driving is an impediment to vehicle operation that can increase the risk of road traffic accidents.

Studies have found vastly different relative risks (RR). Two separate studies using case-crossover analysis each calculated RR at 4, while an epidemiological cohort study found RR, when adjusted for crash-risk exposure, of 1.11 for men and 1.21 for women.

A simulation study from the University of Utah Professor David Strayer compared drivers with a blood alcohol content of 0.08% to those conversing on a cell phone, and after controlling for driving difficulty and time on task, the study concluded that cell phone drivers exhibited greater impairment than intoxicated drivers. Meta-analysis by The Canadian Automobile Association and The University of Illinois found that response time while using both hands-free and hand-held phones was approximately 0.5 standard deviations higher than normal driving (i.e., an average driver, while talking on a cell phone, has response times of a driver in roughly the 40th percentile).

Driving while using a hands-free device is not safer than driving while using a hand-held phone, as concluded by case-crossover studies. epidemiological studies, simulation studies, and meta-analysis. Even with this information, California initiated new Wireless Communications Device Law (effective January 1, 2009) makes it an infraction to write, send, or read text-based communication on an electronic wireless communications device, such as a cell phone, while driving a motor vehicle. Two additional laws dealing with the use of wireless telephones while driving went into effect July 1, 2008. The first law prohibits all drivers from using a handheld wireless telephone while operating a motor vehicle. The law allows a driver to use a wireless telephone to make emergency calls to a law enforcement agency, a medical provider, the fire department, or other emergency services agency. The base fine for the FIRST offense is $20 and $50 for subsequent convictions. With penalty assessments, the fine can be more than triple the base fine amount. videos about California cellular phone laws; with captions (California Vehicle Code [VC] §23123). Motorists 18 and over may use a "hands-free device. The second law effective July 1, 2008, prohibits drivers under the age of 18 from using a wireless telephone or hands-free device while operating a motor vehicle (VC §23124)The consistency of increased crash risk between hands-free and hand-held phone use is at odds with legislation in over 30 countries that prohibit hand-held phone use but allow hands-free. Scientific literature is mixed on the dangers of talking on a phone versus those of talking with a passenger, with the Accident Research Unit at the University of Nottingham finding that the number of utterances was usually higher for mobile calls when compared to blindfolded and non-blindfolded passengers, but the University of Illinois meta-analysis concluding that passenger conversations were just as costly to driving performance as cell phone ones.

Use on Aircraft

As of 2007, several airlines are experimenting with base station and antenna systems installed on the airplane, allowing low power, short-range connection of any phones aboard to remain connected to the aircraft's base station. Thus, they would not attempt connection to the ground base stations as during take off and landing. Simultaneously, airlines may offer phone services to their travelling passengers either as full voice and data services, or initially only as SMS text messaging and similar services. The Australian airline Qantas is the first airline to run a test aeroplane in this configuration in the autumn of 2007. Emirates has announced plans to allow limited mobile phone usage on some flights. However, in the past, commercial airlines have prevented the use of cell phones and laptops, due to the assertion that the frequencies emitted from these devices may disturb the radio waves contact of the airplane.

On March 20, 2008, an Emirates flight was the first time voice calls have been allowed in-flight on commercial airline flights. The breakthrough came after the European Aviation Safety Agency (EASA) and the United Arab Emirates-based General Civil Aviation Authority (GCAA) granted full approval for the AeroMobile system to be used on Emirates. Passengers were able to make and receive voice calls as well as use text messaging. The system automatically came into operation as the Airbus A340-300 reached cruise altitude. Passengers wanting to use the service received a text message welcoming them to the AeroMobile system when they first switched their phones on. The approval by EASA has established that GSM phones are safe to use on airplanes, as the AeroMobile system does not require the modification of aircraft components deemed "sensitive," nor does it require the use of modified phones.

In any case, there are inconsistencies between practices allowed by different airlines and even on the same airline in different countries. For example, Delta Air Lines may allow the use of mobile phones immediately after landing on a domestic flight within the US, whereas they may state "not until the doors are open" on an international flight arriving in the Netherlands. In April 2007 the US Federal Communications Commission officially prohibited passengers' use of cell phones during a flight.

In a similar vein, signs are put up in many countries, such as Canada, the UK and the U.S., at petrol stations prohibiting the use of mobile phones, due to possible safety issues. However, it is unlikely that mobile phone use can cause any problems, and in fact "petrol station employees have themselves spread the rumour about alleged incidents."

Environmental Impacts

Cellular antenna disguised to look like a tree

Like all high structures, cellular antenna masts pose a hazard to low flying aircraft. Towers over a certain height or towers that are close to airports or heliports are normally required to have warning lights. There have been reports that warning lights on cellular masts, TV-towers and other high structures can attract and confuse birds. US authorities estimate that millions of birds are killed near communication towers in the country each year.

Some cellular antenna towers have been camouflaged to make them less obvious on the horizon, and make them look more like a tree.

An example of the way mobile phones and mobile networks have sometimes been perceived as a threat is the widely reported and later discredited claim that mobile phone masts are associated with the Colony Collapse Disorder (CCD) which has reduced bee hive numbers by up to 75% in many areas, especially near cities in the US. The Independent newspaper cited a scientific study claiming it provided evidence for the theory that mobile phone masts *are* a major cause in the collapse of bee populations, with controlled experiments demonstrating a rapid and catastrophic effect on individual hives near masts. Mobile phones were in fact not covered in the study, and the original researchers have since emphatically disavowed any connection between their research, mobile phones, and CCD, specifically indicating that the Independent article had misinterpreted their results and created "a horror story". While the initial claim of damage to bees was widely reported, the corrections to the story were almost non-existent in the media.

There are more than 500 million used mobile phones in the US sitting on shelves or in landfills, and it is estimated that over 125 million will be discarded this year alone. The problem is growing at a rate of more than two million phones per week, putting tons of toxic waste into landfills daily. Several companies offer to buy back and recycle mobile phones from users. In the United States many unwanted but working mobile phones are donated to women's shelters to allow emergency communication.

Tariff Models

Mobile phone shop in Uganda

Payment Methods

There are two principal ways to pay for mobile telephony: the 'pay-as-you-go' model where conversation time is purchased and added to a phone unit via an Internet account or in shops or ATMs, or the contract model where bills are paid by regular intervals after the service has been consumed. It is increasingly common for a consumer to purchase a basic package and then bolt-on services and functionality to create a subscription customised to the users needs.

Pay as you go (also known as "pre-pay" or "prepaid") accounts were invented simultaneously in Portugal and Italy and today form more than half of all mobile phone subscriptions. USA, Canada, Costa Rica, Japan, Israel and Finland are among the rare countries left where most phones are still contract-based.

Incoming Call Charges

In the early days of mobile telephony, the operators (carriers) charged for all air time consumed by the mobile phone user, which included both outbound and inbound telephone calls. As mobile phone adoption rates increased, competition between operators meant that some decided not to charge for incoming calls in some markets (also called "calling party pays").

The European market adopted a calling party pays model throughout the GSM environment and soon various other GSM markets also started to emulate this model.

In Hong Kong, Singapore, Canada, and the United States, it is common for the party receiving the call to be charged per minute, although a few carriers are beginning to offer unlimited received phone calls. This is called the "Receiving Party Pays" model. In China, it was reported that both of its two operators will adopt the caller-pays approach as early as January 2007.

One disadvantage of the receiving party pays systems is that phone owners keep their phones turned off to avoid receiving unwanted calls, which results in the total voice usage rates (and profits) in Calling Party Pays countries outperform those in Receiving Party Pays countries. To avoid the problem of users keeping their phone turned off, most Receiving Party Pays countries have either switched to Calling Party Pays, or their carriers offer additional incentives such as a large number of monthly minutes at a sufficiently discounted rate to compensate for the inconvenience.

Note that when a user roaming in another country, international roaming tariffs apply to all calls received, regardless of the model adopted in the home country.

Bluetooth

A Bluetooth technology is a high speed low powered wireless technology link that is designed to connect phones or other portable equipment together. It is a specification (IEEE 802.15.1) for the use of low power radio communications to link phones, computers and other network devices over short distance without wires. Wireless signals transmitted with Bluetooth cover short distances, typically up to 30 feet (10 meters).

It is achieved by embedded low cost transceivers into the devices. It supports on the frequency band of 2.45GHz and can support upto 721KBps along with three voice channels. This frequency band has been set aside by international agreement for the use of industrial, scientific and medical devices (ISM).rd-compatible with 1.0 devices.

Bluetooth can connect up to "eight devices" simultaneously and each device offers a unique 48 bit address from the IEEE 802 standard with the connections being made point to point or multipoint.

Now a given feature on everything from smartphones to in-vehicle entertainment systems, Bluetooth has an interesting history and working, proving how versatile this communication standard is. It is managed by the Bluetooth Special Interest Group (SIG), a non-profit, non stock company which mainly deals with setting standards, licensing and advancing Bluetooth capabilities. Before this it was standardized by the IEEE as IEEE 802.15.1, but since the SIG, that convention is not used.

Originally conceived as an alternative to the popular RS-232 cables, Bluetooth was invented by telecom giant Ericsson in 1994. Supported by Intel, it then created the above mentioned body called the Bluetooth Special Interest Group that oversaw all development and licensing. For any company to use the standard or market products with the technology, they had to become a member of the SIG. As a result, the SIG is today over 20,000 members strong! However, not all members have the same power, there is a division called 'Promoter Members' for the 'elite members', so to speak. Members from this group have a say in the direction the company is going, and have generally had a hand in the development of the standard as a whole. The Promoter members are:

- Ericsson
- Intel
- Microsoft
- Nokia
- Lenovo
- Toshiba
- Motorola

The Bluetooth Special Interest Group has a set of terms and conditions that must be followed by all members, and also has a set of compliance guidelines for all devices.

Now, to the very interesting name. Ericsson, the company that started Bluetooth is from Sweden, which is part of the Scandinavian region, a historical and cultural-linguistic part of Europe. The name comes from an epithet of a tenth-century King of Denmark and Norway named Harald "Bluetooth" Gormsson. In the local language, he was called *Blåtand* or *Blåtann,* which translated in English became 'Bluetooth'. He was known for uniting the Vikings in ages past, from which the idea of the communication standard came, something that was a single unifying standard for mobile technologies. The logo, in fact, is a combination of two Nordic runes called 'Hagall' and 'Bjarkan', which were the initials of King Harald "Bluetooth" Gormsson.

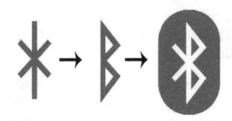

Implementation

Bluetooth operates at frequencies between 2402 and 2480 MHz, or 2400 and 2483.5 MHz including guard bands 2 MHz wide at the bottom end and 3.5 MHz wide at the top. This is in the globally unlicensed (but not unregulated) industrial, scientific and medical (ISM) 2.4 GHz short-range radio frequency band. Bluetooth uses a radio technology called frequency-hopping spread spectrum. Bluetooth divides transmitted data into packets, and transmits each packet on one of 79 designated Bluetooth channels. Each channel has a bandwidth of 1 MHz. It usually performs 800 hops per second, with Adaptive Frequency-Hopping (AFH) enabled. Bluetooth Low Energy uses 2 MHz spacing, which accommodates 40 channels.

Originally, Gaussian frequency-shift keying (GFSK) modulation was the only modulation scheme available. Since the introduction of Bluetooth 2.0+EDR, $\pi/4$-DQPSK (differential quadrature phase shift keying) and 8DPSK modulation may also be used between compatible devices. Devices functioning with GFSK are said to be operating in basic rate (BR) mode where an instantaneous bit rate of 1 Mbit/s is possible. The term Enhanced Data Rate (EDR) is used to describe $\pi/4$-DPSK and 8DPSK schemes, each giving 2 and 3 Mbit/s respectively. The combination of these (BR and EDR) modes in Bluetooth radio technology is classified as a "BR/EDR radio".

Bluetooth is a packet-based protocol with a master/slave architecture. One master may communicate with up to seven slaves in a piconet. All devices share the master's clock. Packet exchange is based on the basic clock, defined by the master, which ticks at 312.5 μs intervals. Two clock ticks make up a slot of 625 μs, and two slots make up a slot pair of 1250 μs. In the simple case of single-slot packets the master transmits in even slots and receives in odd slots. The slave, conversely, receives in even slots and transmits in odd slots. Packets may be 1, 3 or 5 slots long, but in all cases the master's transmission begins in even slots and the slave's in odd slots.

The above is valid for "classic" BT. Bluetooth Low Energy, introduced in the 4.0 specification, uses the same spectrum but somewhat differently.

Communication and Connection

A master BR/EDR Bluetooth device can communicate with a maximum of seven devices in a piconet (an ad-hoc computer network using Bluetooth technology), though not all devices reach this maximum. The devices can switch roles, by agreement, and the slave can become the master (for example, a headset initiating a connection to a phone necessarily begins as master—as initiator of the connection—but may subsequently operate as slave).

The Bluetooth Core Specification provides for the connection of two or more piconets to form a scatternet, in which certain devices simultaneously play the master role in one piconet and the slave role in another.

At any given time, data can be transferred between the master and one other device (except for the little-used broadcast mode). The master chooses which slave device to address; typically, it switches rapidly from one device to another in a round-robin fashion. Since it is the master that chooses which slave to address, whereas a slave is (in theory) supposed to listen in each receive slot, being a master is a lighter burden than being a slave. Being a master of seven slaves is possible; being a slave of more than one master is possible. The specification is vague as to required behavior in scatternets.

Uses

Ranges of Bluetooth devices by class			
Class	Max. permitted power		Typ. range (m)
	(mW)	(dBm)	
1	100	20	~100
2	2.5	4	~10
3	1	0	~1
4	0.5	−3	~0.5

Bluetooth is a standard wire-replacement communications protocol primarily designed for low-power consumption, with a short range based on low-cost transceiver microchips in each device. Because the devices use a radio (broadcast) communications system, they do not have to be in visual line of sight of each other; however, a *quasi optical* wireless path must be viable. Range is power-class-dependent, but effective ranges vary in practice.

Officially Class 3 radios have a range of up to 1 metre (3 ft), Class 2, most commonly found in mobile devices, 10 metres (33 ft), and Class 1, primarily for industrial use cases,100 metres (300 ft). Bluetooth Marketing qualifies that Class 1 range is in most cases 20–30 metres (66–98 ft), and Class 2 range 5–10 metres (16–33 ft). The actual range achieved by a given link will depend on the qualities of the devices at both ends of the link, as well as the air conditions in between, and other factors.

The effective range varies depending on propagation conditions, material coverage, production sample variations, antenna configurations and battery conditions. Most Bluetooth applications are for indoor conditions, where attenuation of walls and signal fading due to signal reflections make the range far lower than specified line-of-sight ranges of the Bluetooth products.

Most Bluetooth applications are battery-powered Class 2 devices, with little difference in range whether the other end of the link is a Class 1 or Class 2 device as the lower-powered device tends to set the range limit. In some cases the effective range of the data link can be extended when a Class 2 device is connecting to a Class 1 transceiver with both higher sensitivity and transmission power than a typical Class 2 device. Mostly, however, the Class 1 devices have a similar sensitivity to Class 2 devices. Connecting two Class 1 devices with both high sensitivity and high power can allow ranges far in excess of the typical 100m, depending on the throughput required by the application. Some such devices allow open field ranges of up to 1 km and beyond between two similar devices without exceeding legal emission limits.

The Bluetooth Core Specification mandates a range of not less than 10 metres (33 ft), but there is no upper limit on actual range. Manufacturers' implementations can be tuned to provide the range needed for each case.

Bluetooth Profile

To use Bluetooth wireless technology, a device must be able to interpret certain Bluetooth profiles, which are definitions of possible applications and specify general behaviors that Bluetooth-enabled devices use to communicate with other Bluetooth devices. These profiles include settings to parameterize and to control the communication from the start. Adherence to profiles saves the time for transmitting the parameters anew before the bi-directional link becomes effective. There are a wide range of Bluetooth profiles that describe many different types of applications or use cases for devices.

List of Applications

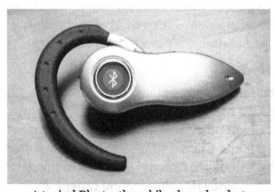

A typical Bluetooth mobile phone headset

- Wireless control of and communication between a mobile phone and a hands-free headset. This was one of the earliest applications to become popular.

- Wireless control of and communication between a mobile phone and a Bluetooth compatible car stereo system.

- Wireless control of and communication with iOS and Android device phones, tablets and portable wireless speakers.

- Wireless Bluetooth headset and Intercom. Idiomatically, a headset is sometimes called "a Bluetooth".

- Wireless streaming of audio to headphones with or without communication capabilities.

- Wireless streaming of data collected by Bluetooth-enabled fitness devices to phone or PC.

- Wireless networking between PCs in a confined space and where little bandwidth is required.

- Wireless communication with PC input and output devices, the most common being the mouse, keyboard and printer.

- Transfer of files, contact details, calendar appointments, and reminders between devices with OBEX.

- Replacement of previous wired RS-232 serial communications in test equipment, GPS receivers, medical equipment, bar code scanners, and traffic control devices.

- For controls where infrared was often used.

- For low bandwidth applications where higher USB bandwidth is not required and cable-free connection desired.

- Sending small advertisements from Bluetooth-enabled advertising hoardings to other, discoverable, Bluetooth devices.

- Wireless bridge between two Industrial Ethernet (e.g., PROFINET) networks.

- Seventh and eighth generation game consoles such as Nintendo's Wii, and Sony's PlayStation 3 use Bluetooth for their respective wireless controllers.

- Dial-up internet access on personal computers or PDAs using a data-capable mobile phone as a wireless modem.

- Short-range transmission of health sensor data from medical devices to mobile phone, set-top box or dedicated telehealth devices.

- Allowing a DECT phone to ring and answer calls on behalf of a nearby mobile phone.

- Real-time location systems (RTLS) are used to track and identify the location of objects in real time using "Nodes" or "tags" attached to, or embedded in, the objects tracked, and "Readers" that receive and process the wireless signals from these tags to determine their locations.

- Personal security application on mobile phones for prevention of theft or loss of items. The protected item has a Bluetooth marker (e.g., a tag) that is in constant communication with the phone. If the connection is broken (the marker is out of range of the phone) then an alarm is raised. This can also be used as a man overboard alarm. A product using this technology has been available since 2009.

- Calgary, Alberta, Canada's Roads Traffic division uses data collected from travelers' Bluetooth devices to predict travel times and road congestion for motorists.

- Wireless transmission of audio (a more reliable alternative to FM transmitters)

- Live video streaming to the visual cortical implant device by Nabeel Fattah in Newcastle university 2017.

Bluetooth vs. Wi-Fi (IEEE 802.11)

Bluetooth and Wi-Fi (Wi-Fi is the brand name for products using IEEE 802.11 standards) have some similar applications: setting up networks, printing, or transferring files. Wi-Fi is intended as a replacement for high-speed cabling for general local area network access in work areas or home. This category of applications is sometimes called wireless local area networks (WLAN). Bluetooth was intended for portable equipment and its applications. The category of applications is outlined as the wireless personal area network (WPAN). Bluetooth is a replacement for cabling in a variety of personally carried applications in any setting, and also works for fixed location applications such as smart energy functionality in the home (thermostats, etc.).

Wi-Fi and Bluetooth are to some extent complementary in their applications and usage. Wi-Fi is usually access point-centered, with an asymmetrical client-server connection with all traffic routed through the access point, while Bluetooth is usually symmetrical, between two Bluetooth devices. Bluetooth serves well in simple applications where two devices need to connect with minimal configuration like a button press, as in headsets and remote controls, while Wi-Fi suits better in applications where some degree of client configuration is possible and high speeds are required, especially for network access through an access node. However, Bluetooth access points do exist, and ad-hoc connections are possible with Wi-Fi though not as simply as with Bluetooth. Wi-Fi Direct was recently developed to add a more Bluetooth-like ad-hoc functionality to Wi-Fi.

Devices

A Bluetooth USB dongle with a 100 m range.

Bluetooth exists in many products, such as telephones, speakers, tablets, media players, robotics systems, handheld, laptops, console gaming equipment as well as some high definition headsets, modems, and watches. The technology is useful when transferring information between two or more devices that are near each other in low-bandwidth situations. Bluetooth is commonly used to transfer sound data with telephones (i.e., with a Bluetooth headset) or byte data with hand-held computers (transferring files).

Bluetooth protocols simplify the discovery and setup of services between devices. Bluetooth devices can advertise all of the services they provide. This makes using services easier, because more of the security, network address and permission configuration can be automated than with many other network types.

Computer Requirements

A personal computer that does not have embedded Bluetooth can use a Bluetooth adapter that enables the PC to communicate with Bluetooth devices. While some desktop computers and most recent laptops come with a built-in Bluetooth radio, others require an external adapter, typically in the form of a small USB "dongle."

A typical Bluetooth USB dongle.

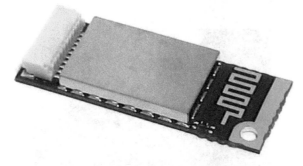

An internal notebook Bluetooth card (14×36×4 mm).

Unlike its predecessor, IrDA, which requires a separate adapter for each device, Bluetooth lets multiple devices communicate with a computer over a single adapter.

Operating System Implementation

For Microsoft platforms, Windows XP Service Pack 2 and SP3 releases work natively with Bluetooth v1.1, v2.0 and v2.0+EDR. Previous versions required users to install their Bluetooth adapter's own drivers, which were not directly supported by Microsoft. Microsoft's own Bluetooth dongles (packaged with their Bluetooth computer devices) have no external drivers and thus require at least Windows XP Service Pack 2. Windows Vista RTM/SP1 with the Feature Pack for Wireless or Windows Vista SP2 work with Bluetooth v2.1+EDR. Windows 7 works with Bluetooth v2.1+EDR and Extended Inquiry Response (EIR).

The Windows XP and Windows Vista/Windows 7 Bluetooth stacks support the following Bluetooth profiles natively: PAN, SPP, DUN, HID, HCRP. The Windows XP stack can be replaced by a third party stack that supports more profiles or newer Bluetooth versions. The Windows Vista/Windows 7 Bluetooth stack supports vendor-supplied additional profiles without requiring that the Microsoft stack be replaced.

Apple products have worked with Bluetooth since [[Mac OS X v10.2]], which was released in 2002.

Linux has two popular Bluetooth stacks, BlueZ and Affix. The BlueZ stack is included with most Linux kernels and was originally developed by Qualcomm. The Affix stack was developed by Nokia.

FreeBSD features Bluetooth since its v5.0 release.

NetBSD features Bluetooth since its v4.0 release. Its Bluetooth stack was ported to OpenBSD as well, however OpenBSD later removed it for security reasons.

Specifications and Features

The specifications were formalized by the Bluetooth Special Interest Group (SIG).

The SIG was formally announced on 20 May 1998. Today it has a membership of over 30,000 companies worldwide. It was established by Ericsson, IBM, Intel, Toshiba and Nokia, and later joined by many other companies.

All versions of the Bluetooth standards support downward compatibility. That lets the latest standard cover all older versions.

The Bluetooth Core Specification Working Group (CSWG) produces mainly 4 kinds of specifications:

- The Bluetooth Core Specification, release cycle is typically a few years in between
- Core Specification Addendum (CSA), release cycle can be as tight as a few times per year
- Core Specification Supplements (CSS), can be released very quickly
- Errata (Available with a user account: Errata login)

Bluetooth 1.0 and 1.0B

Versions 1.0 and 1.0B had many problems, and manufacturers had difficulty making their products interoperable. Versions 1.0 and 1.0B also included mandatory Bluetooth hardware device address (BD_ADDR) transmission in the Connecting process (rendering anonymity impossible at the protocol level), which was a major setback for certain services planned for use in Bluetooth environments.

Bluetooth 1.1

- Ratified as IEEE Standard 802.15.1–2002
- Many errors found in the v1.0B specifications were fixed.
- Added possibility of non-encrypted channels.
- Received Signal Strength Indicator (RSSI).

Bluetooth 1.2

Major enhancements include:

- Faster Connection and Discovery
- *Adaptive frequency-hopping spread spectrum (AFH)*, which improves resistance to radio frequency interference by avoiding the use of crowded frequencies in the hopping sequence.
- Higher transmission speeds in practice than in v1.1, up to 721 kbit/s.

- Extended Synchronous Connections (eSCO), which improve voice quality of audio links by allowing retransmissions of corrupted packets, and may optionally increase audio latency to provide better concurrent data transfer.

- Host Controller Interface (HCI) operation with three-wire UART.

- Ratified as IEEE Standard 802.15.1–2005

- Introduced Flow Control and Retransmission Modes for L2CAP.

Bluetooth 2.0 + EDR

This version of the Bluetooth Core Specification was released in 2004. The main difference is the introduction of an Enhanced Data Rate (EDR) for faster data transfer. The bit rate of EDR is 3 Mbit/s, although the maximum data transfer rate (allowing for inter-packet time and acknowledgements) is 2.1 Mbit/s. EDR uses a combination of GFSK and Phase Shift Keying modulation (PSK) with two variants, $\pi/4$-DQPSK and 8DPSK. EDR can provide a lower power consumption through a reduced duty cycle.

The specification is published as *Bluetooth v2.0 + EDR*, which implies that EDR is an optional feature. Aside from EDR, the v2.0 specification contains other minor improvements, and products may claim compliance to "Bluetooth v2.0" without supporting the higher data rate. At least one commercial device states "Bluetooth v2.0 without EDR" on its data sheet.

Bluetooth 2.1 + EDR

Bluetooth Core Specification Version 2.1 + EDR was adopted by the Bluetooth SIG on 26 July 2007.

The headline feature of v2.1 is secure simple pairing (SSP): this improves the pairing experience for Bluetooth devices, while increasing the use and strength of security.

Version 2.1 allows various other improvements, including "Extended inquiry response" (EIR), which provides more information during the inquiry procedure to allow better filtering of devices before connection; and sniff subrating, which reduces the power consumption in low-power mode.

Bluetooth 3.0 + HS

Version 3.0 + HS of the Bluetooth Core Specification was adopted by the Bluetooth SIG on 21 April 2009. Bluetooth v3.0 + HS provides theoretical data transfer speeds of up to 24 Mbit/s, though not over the Bluetooth link itself. Instead, the Bluetooth link is used for negotiation and establishment, and the high data rate traffic is carried over a colocated 802.11 link.

The main new feature is AMP (Alternative MAC/PHY), the addition of 802.11 as a high-speed transport. The high-speed part of the specification is not mandatory, and hence only devices that display the "+HS" logo actually support Bluetooth over 802.11 high-speed data transfer. A Bluetooth v3.0 device without the "+HS" suffix is only required to support features introduced in Core Specification Version 3.0 or earlier Core Specification Addendum 1.

L2CAP Enhanced modes

Enhanced Retransmission Mode (ERTM) implements reliable L2CAP channel, while Streaming Mode (SM) implements unreliable channel with no retransmission or flow control. Introduced in Core Specification Addendum 1.

Alternative MAC/PHY

Enables the use of alternative MAC and PHYs for transporting Bluetooth profile data. The Bluetooth radio is still used for device discovery, initial connection and profile configuration. However, when large quantities of data must be sent, the high-speed alternative MAC PHY 802.11 (typically associated with Wi-Fi) transports the data. This means that Bluetooth uses proven low power connection models when the system is idle, and the faster radio when it must send large quantities of data. AMP links require enhanced L2CAP modes.

Unicast Connectionless Data

Permits sending service data without establishing an explicit L2CAP channel. It is intended for use by applications that require low latency between user action and reconnection/transmission of data. This is only appropriate for small amounts of data.

Enhanced Power Control

Updates the power control feature to remove the open loop power control, and also to clarify ambiguities in power control introduced by the new modulation schemes added for EDR. Enhanced power control removes the ambiguities by specifying the behaviour that is expected. The feature also adds closed loop power control, meaning RSSI filtering can start as the response is received. Additionally, a "go straight to maximum power" request has been introduced. This is expected to deal with the headset link loss issue typically observed when a user puts their phone into a pocket on the opposite side to the headset.

Ultra-wideband

The high-speed (AMP) feature of Bluetooth v3.0 was originally intended for UWB, but the WiMedia Alliance, the body responsible for the flavor of UWB intended for

Bluetooth, announced in March 2009 that it was disbanding, and ultimately UWB was omitted from the Core v3.0 specification.

On 16 March 2009, the WiMedia Alliance announced it was entering into technology transfer agreements for the WiMedia Ultra-wideband (UWB) specifications. WiMedia has transferred all current and future specifications, including work on future high-speed and power-optimized implementations, to the Bluetooth Special Interest Group (SIG), Wireless USB Promoter Group and the USB Implementers Forum. After successful completion of the technology transfer, marketing, and related administrative items, the WiMedia Alliance ceased operations.

In October 2009 the Bluetooth Special Interest Group suspended development of UWB as part of the alternative MAC/PHY, Bluetooth v3.0 + HS solution. A small, but significant, number of former WiMedia members had not and would not sign up to the necessary agreements for the IP transfer. The Bluetooth SIG is now in the process of evaluating other options for its longer term roadmap.

Bluetooth 4.0 + LE

The Bluetooth SIG completed the Bluetooth Core Specification version 4.0 (called Bluetooth Smart) and has been adopted as of 30 June 2010. It includes *Classic Bluetooth*, *Bluetooth high speed* and *Bluetooth Low Energy* protocols. Bluetooth high speed is based on Wi-Fi, and Classic Bluetooth consists of legacy Bluetooth protocols.

Bluetooth Low Energy, previously known as Wibree, is a subset of Bluetooth v4.0 with an entirely new protocol stack for rapid build-up of simple links. As an alternative to the Bluetooth standard protocols that were introduced in Bluetooth v1.0 to v3.0, it is aimed at very low power applications running off a coin cell. Chip designs allow for two types of implementation, dual-mode, single-mode and enhanced past versions. The provisional names *Wibree* and *Bluetooth ULP* (Ultra Low Power) were abandoned and the BLE name was used for a while. In late 2011, new logos "Bluetooth Smart Ready" for hosts and "Bluetooth Smart" for sensors were introduced as the general-public face of BLE.

Compared to *Classic Bluetooth*, Bluetooth Low Energy is intended to provide considerably reduced power consumption and cost while maintaining a similar communication range. In terms of lengthening the battery life of Bluetooth devices, BLE represents a significant progression.

- In a single-mode implementation, only the low energy protocol stack is implemented. Dialog Semiconductor, STMicroelectronics, AMICCOM, CSR, Nordic Semiconductor and Texas Instruments have released single mode Bluetooth Low Energy solutions.

- In a dual-mode implementation, Bluetooth Smart functionality is integrated into an existing Classic Bluetooth controller. As of March 2011, the following semiconductor companies have announced the availability of chips meeting the standard: Qualcomm-Atheros, CSR, Broadcom and Texas Instruments. The compliant architecture shares all of Classic Bluetooth's existing radio and functionality resulting in a negligible cost increase compared to Classic Bluetooth.

Cost-reduced single-mode chips, which enable highly integrated and compact devices, feature a lightweight Link Layer providing ultra-low power idle mode operation, simple device discovery, and reliable point-to-multipoint data transfer with advanced power-save and secure encrypted connections at the lowest possible cost.

General improvements in version 4.0 include the changes necessary to facilitate BLE modes, as well the Generic Attribute Profile (GATT) and Security Manager (SM) services with AES Encryption.

Core Specification Addendum 2 was unveiled in December 2011; it contains improvements to the audio Host Controller Interface and to the High Speed (802.11) Protocol Adaptation Layer.

Core Specification Addendum 3 revision 2 has an adoption date of 24 July 2012.

Core Specification Addendum 4 has an adoption date of 12 February 2013.

Bluetooth 4.1

The Bluetooth SIG announced formal adoption of the Bluetooth v4.1 specification on 4 December 2013. This specification is an incremental software update to Bluetooth Specification v4.0, and not a hardware update. The update incorporates Bluetooth Core Specification Addenda (CSA 1, 2, 3 & 4) and adds new features that improve consumer usability. These include increased co-existence support for LTE, bulk data exchange rates—and aid developer innovation by allowing devices to support multiple roles simultaneously.

New features of this specification include:

- Mobile Wireless Service Coexistence Signaling
- Train Nudging and Generalized Interlaced Scanning
- Low Duty Cycle Directed Advertising
- L2CAP Connection Oriented and Dedicated Channels with Credit Based Flow Control
- Dual Mode and Topology
- LE Link Layer Topology
- 802.11n PAL

- Audio Architecture Updates for Wide Band Speech

- Fast Data Advertising Interval

- Limited Discovery Time

Notice that some features were already available in a Core Specification Addendum (CSA) before the release of v4.1.

Bluetooth 4.2

Released on December 2, 2014, it introduces features for the Internet of Things.

The major areas of improvement are:

- Low Energy Secure Connection with Data Packet Length Extension

- Link Layer Privacy with Extended Scanner Filter Policies

- Internet Protocol Support Profile (IPSP) version 6 ready for Bluetooth Smart things to support connected home

Older Bluetooth hardware may receive 4.2 features such as Data Packet Length Extension and improved privacy via firmware updates.

Bluetooth 5

The Bluetooth SIG presented Bluetooth 5 on 16 June 2016. Its new features are mainly focused on emerging Internet of Things technology. The Samsung Galaxy S8 launched with Bluetooth 5 support in April 2017. In September 2017, the iPhone 8, 8 Plus and iPhone X launched with Bluetooth 5 support as well. Apple also integrated 'Bluetooth 5.0' in their new HomePod offering released on February 9, 2018. Marketing drops the point number; so that it is just "Bluetooth 5" (not 5.0 or LE like Bluetooth 4.0). The change is for the sake of "Simplifying our marketing, communicating user benefits more effectively and making it easier to signal significant technology updates to the market."

Bluetooth 5 provides, for BLE, options that can double the speed (2 Mbit/s burst) at the expense of range, or up to fourfold the range at the expense of data rate, and eightfold the data broadcasting capacity of transmissions, by increasing the packet lengths. The increase in transmissions could be important for Internet of Things devices, where many nodes connect throughout a whole house. Bluetooth 5 adds functionality for connection-less services such as location-relevant navigation of low-energy Bluetooth connections.

The major areas of improvement are:

- Slot Availability Mask (SAM)

- 2 Mbit/s PHY for LE

- LE Long Range

- High Duty Cycle Non-Connectable Advertising

- LE Advertising Extensions

- LE Channel Selection Algorithm #2

Features Added in CSA5 – Integrated in v5.0:

- Higher Output Power

The following features were removed in this version of the specification:

- Park State

Technical Information

Architecture

Software

> Seeking to extend the compatibility of Bluetooth devices, the devices that adhere to the standard use as interface between the host device (laptop, phone, etc.) and the Bluetooth device as such (Bluetooth chip) an interface called HCI (Host Controller Interface)

> High-level protocols such as the SDP (Protocol used to find other Bluetooth devices within the communication range, also responsible for detecting the function of devices in range), RFCOMM (Protocol used to emulate serial port connections) and TCS (Telephony control protocol) interact with the baseband controller through the L2CAP Protocol (Logical Link Control and Adaptation Protocol). The L2CAP protocol is responsible for the segmentation and reassembly of the packets

Hardware

> The hardware that makes up the Bluetooth device is made up of two parts. A radio device, responsible for modulating and transmitting the signal; and a digital controller. The digital controller is composed of a CPU, a digital signal processor (DSP) called Link Controller and interfaces with the host device. The Link Controller is responsible for the processing of the baseband and the management of ARQ and physical layer FEC protocols. In addition, it handles the transfer functions (both asynchronous and synchronous), audio coding and data encryption. The CPU of the device is responsible for attending the instructions related to Bluetooth of the host device, in order to simplify its operation. To do this, the CPU runs a software called Link Manager that has the function of communicating with other devices through the LMP protocol.

Bluetooth Protocol Stack

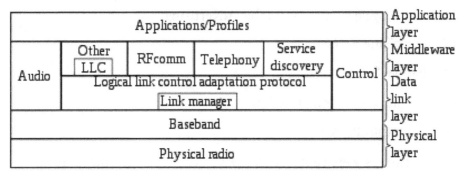

Bluetooth Protocol Stack

Bluetooth is defined as a layer protocol architecture consisting of core protocols, cable replacement protocols, telephony control protocols, and adopted protocols. Mandatory protocols for all Bluetooth stacks are LMP, L2CAP and SDP. In addition, devices that communicate with Bluetooth almost universally can use these protocols: HCI and RFCOMM.

Link Manager

The Link Manager is the system that manages to establish the connection between devices. It is responsible for the establishment, authentication and configuration of the link. The Link Manager locates other managers and communicates with them thanks to the management protocol of the LMP link. In order to perform its function as a service provider, the LM uses the services included in the link controller (LC, "Link Controller"). The Link Manager Protocol basically consists of a number of PDUs (Protocol Data Units) that are sent from one device to another. The following is a list of supported services: • Transmission and reception of data. • Name request • Request of the link addresses. • Establishment of the connection. • Authentication. • Negotiation of link mode and connection establishment.

Host Controller Interface

The Host Controller Interface provides a command interface for the controller and for the link manager, which allows access to the hardware status and control registers. This interface provides an access layer for all devices Bluetooth The HCI layer of the machine exchanges commands and data with the HCI firmware present in the Bluetooth device. One of the most important HCI tasks that must be performed is the automatic discovery of other Bluetooth devices that are within the coverage radius.

Logical Link Control and Adaptation Protocol

The *Logical Link Control and Adaptation Protocol* (L2CAP) is used to multiplex multiple logical connections between two devices using different higher level protocols. Provides segmentation and reassembly of on-air packets.

In *Basic* mode, L2CAP provides packets with a payload configurable up to 64 kB, with 672 bytes as the default MTU, and 48 bytes as the minimum mandatory supported MTU.

In *Retransmission and Flow Control* modes, L2CAP can be configured either for iso-chronous data or reliable data per channel by performing retransmissions and CRC checks.

Bluetooth Core Specification Addendum 1 adds two additional L2CAP modes to the core specification. These modes effectively deprecate original Retransmission and Flow Control modes:

- Enhanced Retransmission Mode (ERTM): This mode is an improved version of the original retransmission mode. This mode provides a reliable L2CAP channel.

- Streaming Mode (SM): This is a very simple mode, with no retransmission or flow control. This mode provides an unreliable L2CAP channel.

Reliability in any of these modes is optionally and/or additionally guaranteed by the lower layer Bluetooth BDR/EDR air interface by configuring the number of retransmissions and flush timeout (time after which the radio flushes packets). In-order sequencing is guaranteed by the lower layer.

Only L2CAP channels configured in ERTM or SM may be operated over AMP logical links.

Service Discovery Protocol

The *Service Discovery Protocol* (SDP) allows a device to discover services offered by other devices, and their associated parameters. For example, when you use a mobile phone with a Bluetooth headset, the phone uses SDP to determine which Bluetooth profiles the headset can use (Headset Profile, Hands Free Profile, Advanced Audio Distribution Profile (A2DP) etc.) and the protocol multiplexer settings needed for the phone to connect to the headset using each of them. Each service is identified by a Universally Unique Identifier (UUID), with official services (Bluetooth profiles) assigned a short form UUID (16 bits rather than the full 128).

Radio Frequency Communications

Radio Frequency Communications (RFCOMM) is a cable replacement protocol used for generating a virtual serial data stream. RFCOMM provides for binary data transport and emulates EIA-232 (formerly RS-232) control signals over the Bluetooth baseband layer, i.e. it is a serial port emulation.

RFCOMM provides a simple reliable data stream to the user, similar to TCP. It is used directly by many telephony related profiles as a carrier for AT commands, as well as being a transport layer for OBEX over Bluetooth.

Many Bluetooth applications use RFCOMM because of its widespread support and publicly available API on most operating systems. Additionally, applications that used a serial port to communicate can be quickly ported to use RFCOMM.

Bluetooth Network Encapsulation Protocol

The *Bluetooth Network Encapsulation Protocol* (BNEP) is used for transferring another protocol stack's data via an L2CAP channel. Its main purpose is the transmission of IP packets in the Personal Area Networking Profile. BNEP performs a similar function to SNAP in Wireless LAN.

Audio/Video Control Transport Protocol

The *Audio/Video Control Transport Protocol* (AVCTP) is used by the remote control profile to transfer AV/C commands over an L2CAP channel. The music control buttons on a stereo headset use this protocol to control the music player.

Audio/Video Distribution Transport Protocol

The *Audio/Video Distribution Transport Protocol* (AVDTP) is used by the advanced audio distribution profile to stream music to stereo headsets over an L2CAP channel intended for video distribution profile in the Bluetooth transmission.

Telephony Control Protocol

The *Telephony Control Protocol – Binary* (TCS BIN) is the bit-oriented protocol that defines the call control signaling for the establishment of voice and data calls between Bluetooth devices. Additionally, "TCS BIN defines mobility management procedures for handling groups of Bluetooth TCS devices."

TCS-BIN is only used by the cordless telephony profile, which failed to attract implementers. As such it is only of historical interest.

Adopted Protocols

Adopted protocols are defined by other standards-making organizations and incorporated into Bluetooth's protocol stack, allowing Bluetooth to code protocols only when necessary. The adopted protocols include:

Point-to-Point Protocol (PPP)

> Internet standard protocol for transporting IP datagrams over a point-to-point link.

TCP/IP/UDP

> Foundation Protocols for TCP/IP protocol suite

Object Exchange Protocol (OBEX)

> Session-layer protocol for the exchange of objects, providing a model for object and operation representation

Wireless Application Environment/Wireless Application Protocol (WAE/WAP)

> WAE specifies an application framework for wireless devices and WAP is an open standard to provide mobile users access to telephony and information services.

Baseband Error Correction

Depending on packet type, individual packets may be protected by error correction, either 1/3 rate forward error correction (FEC) or 2/3 rate. In addition, packets with CRC will be retransmitted until acknowledged by automatic repeat request (ARQ).

Setting up Connections

Any Bluetooth device in *discoverable mode* transmits the following information on demand:

- Device name
- Device class
- List of services
- Technical information (for example: device features, manufacturer, Bluetooth specification used, clock offset)

Any device may perform an inquiry to find other devices to connect to, and any device can be configured to respond to such inquiries. However, if the device trying to connect knows the address of the device, it always responds to direct connection requests and transmits the information shown in the list above if requested. Use of a device's services may require pairing or acceptance by its owner, but the connection itself can be initiated by any device and held until it goes out of range. Some devices can be connected to only one device at a time, and connecting to them prevents them from connecting to other devices and appearing in inquiries until they disconnect from the other device.

Every device has a unique 48-bit address. However, these addresses are generally not shown in inquiries. Instead, friendly Bluetooth names are used, which can be set by the user. This name appears when another user scans for devices and in lists of paired devices.

Most cellular phones have the Bluetooth name set to the manufacturer and model of the phone by default. Most cellular phones and laptops show only the Bluetooth names and special programs are required to get additional information about remote devices. This can be confusing as, for example, there could be several cellular phones in range named T610.

Pairing and Bonding

Motivation

Many services offered over Bluetooth can expose private data or let a connecting party control the Bluetooth device. Security reasons make it necessary to recognize specific devices, and thus enable control over which devices can connect to a given Bluetooth device. At the same time, it is useful for Bluetooth devices to be able to establish a connection without user intervention (for example, as soon as in range).

To resolve this conflict, Bluetooth uses a process called *bonding*, and a bond is generated through a process called *pairing*. The pairing process is triggered either by a specific request from a user to generate a bond (for example, the user explicitly requests to "Add a Bluetooth device"), or it is triggered automatically when connecting to a service where (for the first time) the identity of a device is required for security purposes. These two cases are referred to as dedicated bonding and general bonding respectively.

Pairing often involves some level of user interaction. This user interaction confirms the identity of the devices. When pairing successfully completes, a bond forms between the two devices, enabling those two devices to connect to each other in the future without repeating the pairing process to confirm device identities. When desired, the user can remove the bonding relationship.

Implementation

During pairing, the two devices establish a relationship by creating a shared secret known as a *link key*. If both devices store the same link key, they are said to be *paired* or *bonded*. A device that wants to communicate only with a bonded device can cryptographically authenticate the identity of the other device, ensuring it is the same device it previously paired with. Once a link key is generated, an authenticated Asynchronous Connection-Less (ACL) link between the devices may be encrypted to protect exchanged data against eavesdropping. Users can delete link keys from either device, which removes the bond between the devices—so it is possible for one device to have a stored link key for a device it is no longer paired with.

Bluetooth services generally require either encryption or authentication and as such require pairing before they let a remote device connect. Some services, such as the Object Push Profile, elect not to explicitly require authentication or encryption so that pairing does not interfere with the user experience associated with the service use-cases.

Pairing Mechanisms

Pairing mechanisms changed significantly with the introduction of Secure Simple Pairing in Bluetooth v2.1. The following summarizes the pairing mechanisms:

- *Legacy pairing*: This is the only method available in Bluetooth v2.0 and before. Each device must enter a PIN code; pairing is only successful if both devices enter the same PIN code. Any 16-byte UTF-8 string may be used as a PIN code; however, not all devices may be capable of entering all possible PIN codes.

  o *Limited input devices*: The obvious example of this class of device is a Bluetooth Hands-free headset, which generally have few inputs. These devices usually have a *fixed PIN*, for example "0000" or "1234", that are hard-coded into the device.

  o *Numeric input devices*: Mobile phones are classic examples of these devices. They allow a user to enter a numeric value up to 16 digits in length.

  o *Alpha-numeric input devices*: PCs and smartphones are examples of these devices. They allow a user to enter full UTF-8 text as a PIN code. If pairing with a less capable device the user must be aware of the input limitations on the other device; there is no mechanism available for a capable device to determine how it should limit the available input a user may use.

- *Secure Simple Pairing* (SSP): This is required by Bluetooth v2.1, although a Bluetooth v2.1 device may only use legacy pairing to interoperate with a v2.0 or earlier device. Secure Simple Pairing uses a form of public key cryptography, and some types can help protect against man in the middle, or MITM attacks. SSP has the following authentication mechanisms:

  o *Just works*: As the name implies, this method just works, with no user interaction. However, a device may prompt the user to confirm the pairing process. This method is typically used by headsets with very limited IO capabilities, and is more secure than the fixed PIN mechanism this limited set of devices uses for legacy pairing. This method provides no man-in-the-middle (MITM) protection.

  o *Numeric comparison*: If both devices have a display, and at least one can accept a binary yes/no user input, they may use Numeric Comparison. This method displays a 6-digit numeric code on each device. The user should compare the numbers to ensure they are identical. If the comparison succeeds, the user(s) should confirm pairing on the device(s) that can accept an input. This method provides MITM protection, assuming the user confirms on both devices and actually performs the comparison properly.

- *Passkey Entry*: This method may be used between a device with a display and a device with numeric keypad entry (such as a keyboard), or two devices with numeric keypad entry. In the first case, the display presents a 6-digit numeric code to the user, who then enters the code on the keypad. In the second case, the user of each device enters the same 6-digit number. Both of these cases provide MITM protection.

- *Out of band* (OOB): This method uses an external means of communication, such as near-field communication (NFC) to exchange some information used in the pairing process. Pairing is completed using the Bluetooth radio, but requires information from the OOB mechanism. This provides only the level of MITM protection that is present in the OOB mechanism.

SSP is considered simple for the following reasons:

- In most cases, it does not require a user to generate a passkey.

- For use cases not requiring MITM protection, user interaction can be eliminated.

- For *numeric comparison*, MITM protection can be achieved with a simple equality comparison by the user.

- Using OOB with NFC enables pairing when devices simply get close, rather than requiring a lengthy discovery process.

Security Concerns

Prior to Bluetooth v2.1, encryption is not required and can be turned off at any time. Moreover, the encryption key is only good for approximately 23.5 hours; using a single encryption key longer than this time allows simple XOR attacks to retrieve the encryption key.

- Turning off encryption is required for several normal operations, so it is problematic to detect if encryption is disabled for a valid reason or for a security attack.

Bluetooth v2.1 addresses this in the following ways:

- Encryption is required for all non-SDP (Service Discovery Protocol) connections

- A new Encryption Pause and Resume feature is used for all normal operations that require that encryption be disabled. This enables easy identification of normal operation from security attacks.

- The encryption key must be refreshed before it expires.

Link keys may be stored on the device file system, not on the Bluetooth chip itself. Many Bluetooth chip manufacturers let link keys be stored on the device—however, if the device is removable, this means that the link key moves with the device.

Security

Bluetooth implements confidentiality, authentication and key derivation with custom algorithms based on the SAFER+ block cipher. Bluetooth key generation is generally based on a Bluetooth PIN, which must be entered into both devices. This procedure might be modified if one of the devices has a fixed PIN (e.g., for headsets or similar devices with a restricted user interface). During pairing, an initialization key or master key is generated, using the E22 algorithm. The E0 stream cipher is used for encrypting packets, granting confidentiality, and is based on a shared cryptographic secret, namely a previously generated link key or master key. Those keys, used for subsequent encryption of data sent via the air interface, rely on the Bluetooth PIN, which has been entered into one or both devices.

An overview of Bluetooth vulnerabilities exploits was published in 2007 by Andreas Becker.

In September 2008, the National Institute of Standards and Technology (NIST) published a Guide to Bluetooth Security as a reference for organizations. It describes Bluetooth security capabilities and how to secure Bluetooth technologies effectively. While Bluetooth has its benefits, it is susceptible to denial-of-service attacks, eavesdropping, man-in-the-middle attacks, message modification, and resource misappropriation. Users and organizations must evaluate their acceptable level of risk and incorporate security into the lifecycle of Bluetooth devices. To help mitigate risks, included in the NIST document are security checklists with guidelines and recommendations for creating and maintaining secure Bluetooth piconets, headsets, and smart card readers.

Bluetooth v2.1 – finalized in 2007 with consumer devices first appearing in 2009 – makes significant changes to Bluetooth's security, including pairing.

Bluejacking

Bluejacking is the sending of either a picture or a message from one user to an unsuspecting user through Bluetooth wireless technology. Common applications include short messages, e.g., "You've just been bluejacked!" Bluejacking does not involve the removal or alteration of any data from the device. Bluejacking can also involve taking control of a mobile device wirelessly and phoning a premium rate line, owned by the bluejacker. Security advances have alleviated this issue.

Health Concerns

Bluetooth uses the microwave radio frequency spectrum in the 2.402 GHz to 2.480 GHz range, which is non-ionizing radiation, of similar bandwidth to the one used by wireless and mobile phones. No specific demonstration of harm has been demonstrated up to date, even if wireless transmission has been included by IARC in the possible carcinogen list. Maximum power output from a Bluetooth radio is 100 mW for class 1, 2.5 mW for class 2, and 1 mW for class 3 devices. Even the maximum power output of class 1 is a lower level than the lowest-powered mobile phones. UMTS and W-CDMA output 250 mW, GSM1800/1900 outputs 1000 mW, and GSM850/900 outputs 2000 mW.

Wi-Fi

Wi-Fi is the name of a popular wireless networking technology that uses radio waves to provide wireless high-speed Internet and network connections. A common misconception is that the term Wi-Fi is short for *"wireless fidelity,"* however this is not the case. Wi-Fi is simply a trademarked phrase that means *IEEE 802.11x.*

Wi-Fi networks have no physical wired connection between sender and receiver by using radio frequency (RF) technology -- a frequency within the electromagnetic spectrum associated with radio wave propagation. When an RF current is supplied to an antenna, an electromagnetic field is created that then is able to propagate through space.

The cornerstone of any wireless network is an access point (AP). The primary job of an access point is to broadcast a wireless signal that computers can detect and "tune" into. In order to connect to an access point and join a wireless network, computers and devices must be equipped with wireless network adapters.

The Wi-Fi Alliance, the organization that owns the Wi-Fi registered trademark term specifically defines Wi-Fi as any *"wireless local area network (WLAN) products that are based on the Institute of Electrical and Electronics Engineers' (IEEE) 802.11 standards."*

The old Wi-Fi Alliance logo	
Introduced	September 1998; 19 years ago
Compatible hardware	Personal computers, gaming consoles, televisions, printers, mobile phones

Initially, Wi-Fi was used in place of only the 2.4GHz 802.11b standard, however the Wi-Fi Alliance has expanded the generic use of the Wi-Fi term to include any type of network or WLAN product based on any of the 802.11 standards, including 802.11b, 802.11a, dual-band and so on, in an attempt to stop confusion about wireless LAN interoperability.

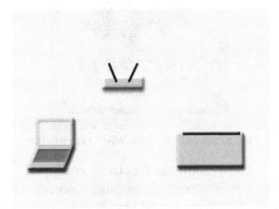

Depiction of a device sending information wirelessly to another device, both connected to the local network, in order to print a document

Wi-Fi Certification

The IEEE does not test equipment for compliance with their standards. The non-profit Wi-Fi Alliance was formed in 1999 to fill this void — to establish and enforce standards for interoperability and backward compatibility, and to promote wireless local-area-network technology. As of 2010, the Wi-Fi Alliance consisted of more than 375 companies from around the world. The Wi-Fi Alliance enforces the use of the Wi-Fi brand to technologies based on the IEEE 802.11 standards from the IEEE. This includes wireless local area network (WLAN) connections, device to device connectivity (such as Wi-Fi Peer to Peer aka Wi-Fi Direct), Personal area network (PAN), local area network (LAN) and even some limited wide area network (WAN) connections. Manufacturers with membership in the Wi-Fi Alliance, whose products pass the certification process, gain the right to mark those products with the Wi-Fi logo.

Specifically, the certification process requires conformance to the IEEE 802.11 radio standards, the WPA2 and WPA3 security standards, and the EAP authentication standard. Certification may optionally include tests of IEEE 802.11 draft standards, interaction with cellular-phone technology in converged devices, and features relating to security set-up, multimedia, and power-saving.

Not every Wi-Fi device is submitted for certification. The lack of Wi-Fi certification does not necessarily imply that a device is incompatible with other Wi-Fi devices. The Wi-Fi Alliance may or may not sanction derivative terms, such as Super Wi-Fi, coined by the US Federal Communications Commission (FCC) to describe proposed networking in the UHF TV band in the US.

Terminology

A Japanese sticker indicating to the public that a location is within range of a Wi-Fi network. A dot with curved lines radiating from it is a common symbol for Wi-Fi, representing a point transmitting a signal.

To connect to a Wi-Fi LAN, a computer has to be equipped with a wireless network interface controller. The combination of computer and interface controllers is called a *station*. For all stations that share a single radio frequency communication channel, transmissions on this channel are received by all stations within range. The transmission is not guaranteed to be delivered and is therefore a best-effort delivery mechanism. A carrier wave is used to transmit the data. The data is organised in packets on an Ethernet link, referred to as "Ethernet frames".

A service set is the set of all the devices associated with a particular Wi-Fi network. The service set can be local, independent, extended or mesh.

Each service set has an associated identifier, the 32-byte Service Set Identifier (SSID), which identifies the particular network. The SSID is configured within the devices that are considered part of the network, and it is transmitted in the packets. Receivers ignore wireless packets from networks with a different SSID.

Uses

Internet Access

Wi-Fi technology may be used to provide Internet access to devices that are within the range of a wireless network that is connected to the Internet. The coverage of one or more interconnected access points (*hotspots*) can extend from an area as small as a few rooms to as large as many square kilometres. Coverage in the larger area may require a group of access points with overlapping coverage. For example, public outdoor Wi-Fi technology has been used successfully in wireless mesh networks in London, UK. An international example is Fon.

Wi-Fi provides service in private homes, businesses, as well as in public spaces at Wi-Fi hotspots set up either free-of-charge or commercially, often using a captive portal webpage for access. Organizations and businesses, such as airports, hotels, and restaurants, often provide free-use hotspots to attract customers. Enthusiasts or authorities who wish to provide services or even to promote business in selected areas sometimes provide free Wi-Fi access.

Routers that incorporate a digital subscriber line modem or a cable modem and a Wi-Fi access point, often set up in homes and other buildings, provide Internet access and internetworking to all devices connected to them, wirelessly or via cable.

Similarly, battery-powered routers may include a cellular Internet radio modem and Wi-Fi access point. When subscribed to a cellular data carrier, they allow nearby Wi-Fi stations to access the Internet over 2G, 3G, or 4G networks using the tethering technique. Many smartphones have a built-in capability of this sort, including those based on Android, BlackBerry, Bada, iOS (iPhone), Windows Phone and Symbian, though carriers often disable the feature, or charge a separate fee to enable it, especially for customers with unlimited data plans. "Internet packs" provide standalone facilities of this type as well, without use of a smartphone; examples include the MiFi- and WiBro-branded devices. Some laptops that have a cellular modem card can also act as mobile Internet Wi-Fi access points.

Wi-Fi also connects places that normally don't have network access, such as kitchens and garden sheds.

Google is intending to use the technology to allow rural areas to enjoy connectivity by utilizing a broad mix of projection and routing services. Google also intends to bring connectivity to Africa and some Asian lands by launching blimps that will allow for internet connection with Wi-Fi technology.

City-wide Wi-Fi

An outdoor Wi-Fi access point

In the early 2000s, many cities around the world announced plans to construct city-wide Wi-Fi networks. There are many successful examples; in 2004, Mysore (Mysuru) became India's first Wi-Fi-enabled city. A company called WiFiyNet has set up hotspots in Mysore, covering the complete city and a few nearby villages.

In 2005, St. Cloud, Florida and Sunnyvale, California, became the first cities in the United States to offer citywide free Wi-Fi (from MetroFi). Minneapolis has generated $1.2 million in profit annually for its provider.

In May 2010, London mayor Boris Johnson pledged to have London-wide Wi-Fi by 2012. Several boroughs including Westminster and Islington already had extensive outdoor Wi-Fi coverage at that point.

Officials in South Korea's capital Seoul are moving to provide free Internet access at more than 10,000 locations around the city, including outdoor public spaces, major streets and densely populated residential areas. Seoul will grant leases to KT, LG Telecom, and SK Telecom. The companies will invest $44 million in the project, which was to be completed in 2015.

Campus-wide Wi-Fi

Many traditional university campuses in the developed world provide at least partial Wi-Fi coverage. Carnegie Mellon University built the first campus-wide wireless Internet network, called Wireless Andrew, at its Pittsburgh campus in 1993 before Wi-Fi branding originated. By February 1997 the CMU Wi-Fi zone was fully operational. Many universities collaborate in providing Wi-Fi access to students and staff through the Eduroam international authentication infrastructure.

Wi-Fi Ad Hoc versus Wi-Fi direct

Wi-Fi also allows communications directly from one computer to another without an access point intermediary. This is called *ad hoc* Wi-Fi transmission. This wireless ad hoc network mode has proven popular with multiplayer handheld game consoles, such as the Nintendo DS, PlayStation Portable, digital cameras, and other consumer electronics devices. Some devices can also share their Internet connection using ad hoc, becoming hotspots or "virtual routers".

A keychain-size Wi-Fi detector

Similarly, the Wi-Fi Alliance promotes the specification Wi-Fi Direct for file transfers and media sharing through a new discovery- and security-methodology. Wi-Fi Direct launched in October 2010.

Another mode of direct communication over Wi-Fi is Tunneled Direct Link Setup (TDLS), which enables two devices on the same Wi-Fi network to communicate directly, instead of via the access point.

Wi-Fi Radio Spectrum

802.11b and 802.11g use the 2.4 GHz ISM band, operating in the United States under Part 15 Rules and Regulations. Because of this choice of frequency band, 802.11b and g equipment may occasionally suffer interference from microwave ovens, cordless telephones, USB3.0 hubs, and Bluetooth devices.

Spectrum assignments and operational limitations are not consistent worldwide: Australia and Europe allow for an additional two channels (12, 13) beyond the 11 permitted in the United States for the 2.4 GHz band, while Japan has three more (12–14). In the US and other countries, 802.11a and 802.11g devices may be operated without a license, as allowed in Part 15 of the FCC Rules and Regulations.

A Wi-Fi signal occupies five channels in the 2.4 GHz band. Any two channel numbers that differ by five or more, such as 2 and 7, do not overlap. The oft-repeated adage that channels 1, 6, and 11 are the *only* non-overlapping channels is, therefore, not accurate. Channels 1, 6, and 11 are the only *group of three* non-overlapping channels in North America and the United Kingdom. In Europe and Japan using Channels 1, 5, 9, and 13 for 802.11g and 802.11n is recommended.

802.11a uses the 5 GHz U-NII band, which, for much of the world, offers at least 23 non-overlapping channels rather than the 2.4 GHz ISM frequency band, where adjacent channels overlap.

Interference

Wi-Fi connections can be disrupted or the Internet speed lowered by having other devices in the same area. Wi-Fi protocols are designed to share channels reasonably fairly, and will often work with little to no disruption. However, many 2.4 GHz 802.11b and 802.11g access-points default to the same channel on initial startup, contributing to congestion on certain channels. Wi-Fi pollution, or an excessive number of access points in the area, can prevent access and interfere with other devices' use of other access points as well as with decreased signal-to-noise ratio (SNR) between access points. In addition interference can be caused by overlapping channels in the 802.11g/b spectrum. These issues can become a problem in high-density areas, such as large apartment complexes or office buildings with many Wi-Fi access points.

Additionally, other devices use the 2.4 GHz band: microwave ovens, ISM band devices, security cameras, ZigBee devices, Bluetooth devices, video senders, cordless phones, baby monitors, and, in some countries, amateur radio, all of which can cause significant additional interference. It is also an issue when municipalities or other large entities (such as universities) seek to provide large area coverage.

These bands are allowed to be used with low power transmitters, without requiring a license and with few restrictions. However, while unintended interference is common, users that have been found to knowingly cause deliberate interference to other users, particularly for attempting to locally monopolise these bands for commercial purposes, have been handed large fines.

Communications Protocol

USB wireless adapter

The IEEE 802.11 standard is a set of media access control (MAC) and physical layer (PHY) specifications for implementing wireless local area network (WLAN) computer communication in the 2.4, 3.6, 5, and 60 GHz frequency bands. They are created and maintained by the IEEE LAN/MAN Standards Committee (IEEE 802). The base version of the standard was released in 1997, and has had subsequent amendments. The standard and amendments provide the basis for wireless network products using the Wi-Fi brand. While each amendment is officially revoked when it is incorporated in the latest version of the standard, the corporate world tends to market to the revisions because they concisely denote capabilities of their products. As a result, in the market place, each revision tends to become its own standard.

Range and Speed

Parabolic dishes transmit and receive the radio waves only in particular directions and can give much greater range than omnidirection antennas

Yagi-Uda antennas, widely used for television reception, are relatively compact at WiFi wavelengths

Wi-Fi operational range depends on factors such as the frequency band, radio power output, receiver sensitivity, antenna gain and antenna type as well as the modulation technique. In addition, propagation characteristics of the signals can have a big impact.

At longer distances, and with greater signal absorption, speed is usually reduced.

Transmitter Power

Compared to cell phones and similar technology, Wi-Fi transmitters are low power devices. In general, the maximum amount of power that a Wi-Fi device can transmit is limited by local regulations, such as FCC Part 15 in the US. Equivalent isotropically radiated power (EIRP) in the European Union is limited to 20 dBm (100 mW).

To reach requirements for wireless LAN applications, Wi-Fi has higher power consumption compared to some other standards designed to support wireless personal area network (PAN) applications. For example Bluetooth provides a much shorter propagation range between 1 and 100 m and so in general have a lower power consumption. Other low-power technologies such as ZigBee have fairly long range, but much lower data rate. The high power consumption of Wi-Fi makes battery life in some mobile devices a concern.

Antenna

An access point compliant with either 802.11b or 802.11g, using the stock omnidirectional antenna might have a range of 100 m (0.062 mi). The same radio with an external semi parabolic antenna (15 dB gain) with a similarly equipped receiver at the far end might have a range over 20 miles.

Higher gain rating (dBi) indicates further deviation (generally toward the horizontal) from a theoretical, perfect isotropic radiator, and therefore the antenna can project a usable signal further in particular directions, as compared to a similar output power on a more isotropic antenna. For example, an 8 dBi antenna used with a 100 mW driver will have a similar horizontal range to a 6 dBi antenna being driven at 500 mW. Note that this assumes that radiation in the vertical is lost; this may not be the case in some situations, especially in large buildings or within a waveguide. In the above example, a directional waveguide could cause the low power 6 dBi antenna to project much further in a single direction than the 8 dBi antenna which is not in a waveguide, even if they are both being driven at 100 mW.

On wireless routers with detachable antennas, it is possible to improve range by fitting upgraded antennas which have higher gain in particular directions. Outdoor ranges can be improved to many kilometers through the use of high gain directional antennas at the router and remote device(s).

MIMO

This Netgear Wi-Fi router contains dual bands for transmitting the
802.11 standard across the 2.4 and 5 GHz spectrums and supports MIMO.

Some standards, such as IEEE 802.11n and IEEE 802.11ac for Wi-Fi allow a device to have multiple antennas. Multiple antennas enable the equipment to focus on the far end device, reducing interference in other directions, and giving a stronger useful signal. This greatly increases range and network speed without exceeding the legal power limits.

IEEE802.11n can more than double the range. Range also varies with frequency band. Wi-Fi in the 2.4 GHz frequency block has slightly better range than Wi-Fi in the 5 GHz frequency block used by 802.11a (and optionally by 802.11n).

Under optimal conditions, IEEE802.11ac can achieve communication rates of 1Gb/s.

Alternatives

For the best performance, a number of people only recommend using wireless networking as a supplement to wired networking.

Researchers have developed a number of "no new wires" technologies to provide alternatives to Wi-Fi for applications in which Wi-Fi's indoor range is not adequate and where installing new wires (such as CAT-6) is not possible or cost-effective. For example, the ITU-T G.hn standard for high speed local area networks uses existing home wiring (coaxial cables, phone lines and power lines). Although G.hn does not provide some of the advantages of Wi-Fi (such as mobility or outdoor use), it is designed for applications (such as IPTV distribution) where indoor range is more important than mobility.

Many mobile devices also employ the cell phone networks to provide access to the Internet when out of range of Wi-Fi networks. This is done, rather than rely entirely on cell networks as Wi-Fi networks are often cheaper for bulk data usage.

Radio Propagation

With Wi-Fi signals line-of-sight usually works best, but Wi-Fi signals can be affected by absorption, reflection, and diffraction through and around structures.

Due to the complex nature of radio propagation at typical Wi-Fi frequencies, particularly the effects of signal reflection off trees and buildings, algorithms can only approximately predict Wi-Fi signal strength for any given area in relation to a transmitter. This effect does not apply equally to long-range Wi-Fi, since longer links typically operate from towers that transmit above the surrounding foliage.

The practical range of Wi-Fi essentially confines mobile use to such applications as inventory-taking machines in warehouses or in retail spaces, barcode-reading devices at check-out stands, or receiving/shipping stations. Mobile use of Wi-Fi over wider ranges is limited, for instance, to uses such as in an automobile moving from one hotspot to another. Other wireless technologies are more suitable for communicating with moving vehicles.

Distance Records

Distance records (using non-standard devices) include 382 km (237 mi) in June 2007, held by Ermanno Pietrosemoli and EsLaRed of Venezuela, transferring about 3 MB of data between the mountain-tops of El Águila and Platillon. The Swedish Space Agency transferred data 420 km (260 mi), using 6 watt amplifiers to reach an overhead stratospheric balloon.

Throughput

As the 802.11 specifications evolved to support higher throughput, the bandwidth requirements also increased to support them. 802.11n uses double the radio spectrum/bandwidth (40 MHz) compared to 802.11a or 802.11g (20 MHz).76 This means there can be only one 802.11n network on the 2.4 GHz band at a given location, without interference to/from other WLAN traffic. 802.11n can also be set to limit itself to 20 MHz bandwidth to prevent interference in dense community.

Many newer consumer devices support the latest 802.11ac standard, which uses the 5 GHz band exclusively and is capable of multi-station WLAN throughput of at least 1 gigabit per second, and a single station throughput of at least 500 Mbit/s. In the first quarter of 2016, The Wi-Fi Alliance certifies devices compliant with the 802.11ac standard as "Wi-Fi CERTIFIED ac". This new standard uses several advanced signal processing techniques such as multi-user MIMO and 4X4 Spatial Multiplexing streams, and large channel bandwidth (160 MHz) to achieve the Gigabit throughput. According to a study by IHS Technology, 70% of all access point sales revenue In the first quarter of 2016 came from 802.11ac devices.

Multiple Access Points

Increasing the number of Wi-Fi access points for a network provides redundancy, better range, support for fast roaming and increased overall network-capacity by using more channels or by defining smaller cells. Except for the smallest implementations (such as home or small office networks), Wi-Fi implementations have moved toward "thin" access points, with more of the network intelligence housed in a centralized network appliance, relegating individual access points to the role of "dumb" transceivers. Outdoor applications may use mesh topologies.

When multiple access points are deployed they are often configured with the same SSID and any security settings to form an "extended service set". Wi-Fi client devices will typically connect to the access point that can provide the strongest signal within that service set.

Hardware

Wi-Fi whitelist triggered on an HP laptop

Wi-Fi allows wireless deployment of local area networks (LANs). Also, spaces where cables cannot be run, such as outdoor areas and historical buildings, can host wireless LANs. However, building walls of certain materials, such as stone with high metal content, can block Wi-Fi signals.

Since early 2000's manufacturers are building wireless network adapters into most laptops. The price of chipsets for Wi-Fi continues to drop, making it an economical networking option included in even more devices.

Different competitive brands of access points and client network-interfaces can inter-operate at a basic level of service. Products designated as "Wi-Fi Certified" by the Wi-Fi Alliance are backward compatible. Unlike mobile phones, any standard Wi-Fi device will work anywhere in the world.

Standard Devices

A wireless access point (WAP) connects a group of wireless devices to an adjacent wired LAN. An access point resembles a network hub, relaying data between connected wireless devices in addition to a (usually) single connected wired device, most often an Ethernet hub or switch, allowing wireless devices to communicate with other wired devices.

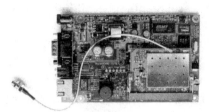  

| An embedded RouterBoard 112 with U.FL-RSMA pigtail and R52 mini PCI Wi-Fi card widely used by wireless Internet service providers (WISPs) in the Czech Republic | OSBRiDGE 3GN – 802.11n Access Point and UMTS/GSM Gateway in one device | An Atheros draft-N Wi-Fi adapter with built in Blue-tooth on a laptop |

Wireless adapters allow devices to connect to a wireless network. These adapters connect to devices using various external or internal interconnects such as PCI, miniPCI, USB, ExpressCard, Cardbus and PC Card. As of 2010, most newer laptop computers come equipped with built in internal adapters.

Wireless routers integrate a Wireless Access Point, Ethernet switch, and internal router firmware application that provides IP routing, NAT, and DNS forwarding through an integrated WAN-interface. A wireless router allows wired and wireless Ethernet LAN devices to connect to a (usually) single WAN device such as a cable modem or a DSL modem. A wireless router allows all three devices, mainly the access point and router, to be configured through one central utility. This utility is usually an integrated web server that is accessible to wired and wireless LAN clients and often optionally to WAN clients. This utility may also be an application that is run on a computer, as is the case with as Apple's AirPort, which is managed with the AirPort Utility on macOS and iOS.

Wireless network bridges connect a wired network to a wireless network. A bridge differs from an access point: an access point connects wireless devices to a wired network at the data-link layer. Two wireless bridges may be used to connect two wired networks over a wireless link, useful in situations where a wired connection may be unavailable, such as between two separate homes or for devices which do not have wireless networking capability (but have wired networking capability), such as consumer entertainment devices; alternatively, a wireless bridge can be used to enable a device which supports a wired connection to operate at a wireless networking standard which is faster than supported by the wireless network connectivity feature (external dongle or inbuilt) supported by the device (e.g. enabling Wireless-N speeds (up to the maximum supported speed on the wired Ethernet port on both the bridge and connected devices including the wireless access point) for a device which only supports Wireless-G). A dual-band wireless bridge can also be used to enable 5 GHz wireless network operation on a device which only supports 2.4 GHz wireless networking functionality and has a wired Ethernet port.

Wireless range-extenders or wireless repeaters can extend the range of an existing wireless network. Strategically placed range-extenders can elongate a signal area or allow for the signal area to reach around barriers such as those pertaining in L-shaped corridors. Wireless devices connected through repeaters will suffer from an increased latency for each hop, as well as from a reduction in the maximum data throughput that is available. In addition, the effect of additional users using a network employing wireless range-extenders is to consume the available bandwidth faster than would be the case where but a single user migrates around a network employing extenders. For this reason, wireless range-extenders work best in networks supporting very low traffic throughput requirements, such as for cases where but a single user with a Wi-Fi equipped tablet migrates around the combined extended and non-extended portions of the total connected network. Additionally, a wireless device connected to any of the repeaters in the chain will have a data throughput that is also limited by the "weakest link" existing in the chain between where the connection originates and where the connection ends. Networks employing wireless extenders are also more prone to degradation from interference from neighboring access points that border portions of the extended network and that happen to occupy the same channel as the extended network.

The security standard, Wi-Fi Protected Setup, allows embedded devices with limited graphical user interface to connect to the Internet with ease. Wi-Fi Protected Setup has 2 configurations: The Push Button configuration and the PIN configuration. These embedded devices are also called The Internet of Things and are low-power, battery-operated embedded systems. A number of Wi-Fi manufacturers design chips and modules for embedded Wi-Fi, such as GainSpan.

Embedded Systems

Embedded serial-to-Wi-Fi module

Increasingly in the last few years (particularly as of 2007), embedded Wi-Fi modules have become available that incorporate a real-time operating system and provide a simple means of wirelessly enabling any device which has and communicates via a serial port. This allows the design of simple monitoring devices. An example is a portable

ECG device monitoring a patient at home. This Wi-Fi-enabled device can communicate via the Internet.

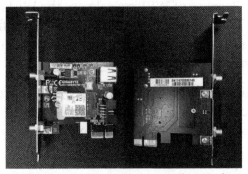

Wireless network interface controller Gigabyte GC-WB867D-I.

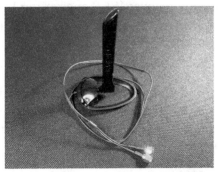

Antenna of wireless network interface controller Gigabyte GC-WB867D-I.

These Wi-Fi modules are designed by OEMs so that implementers need only minimal Wi-Fi knowledge to provide Wi-Fi connectivity for their products.

In June 2014, Texas Instruments introduced the first ARM Cortex-M4 microcontroller with an onboard dedicated Wi-Fi MCU, the SimpleLink CC3200. It makes embedded systems with Wi-Fi connectivity possible to build as single-chip devices, which reduces their cost and minimum size, making it more practical to build wireless-networked controllers into inexpensive ordinary objects.

Network Security

The main issue with wireless network security is its simplified access to the network compared to traditional wired networks such as Ethernet. With wired networking, one must either gain access to a building (physically connecting into the internal network), or break through an external firewall. To enable Wi-Fi, one merely needs to be within the range of the Wi-Fi network. Most business networks protect sensitive data and systems by attempting to disallow external access. Enabling wireless connectivity reduces security if the network uses inadequate or no encryption.

An attacker who has gained access to a Wi-Fi network router can initiate a DNS spoofing attack against any other user of the network by forging a response before the queried DNS server has a chance to reply.

Securing Methods

A common measure to deter unauthorized users involves hiding the access point's name by disabling the SSID broadcast. While effective against the casual user, it is ineffective as a security method because the SSID is broadcast in the clear in response to a client SSID query. Another method is to only allow computers with known MAC addresses to join the network, but determined eavesdroppers may be able to join the network by spoofing an authorized address.

Wired Equivalent Privacy (WEP) encryption was designed to protect against casual snooping but it is no longer considered secure. Tools such as AirSnort or Aircrack-ng can quickly recover WEP encryption keys. Because of WEP's weakness the Wi-Fi Alliance approved Wi-Fi Protected Access (WPA) which uses TKIP. WPA was specifically designed to work with older equipment usually through a firmware upgrade. Though more secure than WEP, WPA has known vulnerabilities.

The more secure WPA2 using Advanced Encryption Standard was introduced in 2004 and is supported by most new Wi-Fi devices. WPA2 is fully compatible with WPA. In 2017 a flaw in the WPA2 protocol was discovered, allowing a key replay attack, known as KRACK.

A flaw in a feature added to Wi-Fi in 2007, called Wi-Fi Protected Setup (WPS), allows WPA and WPA2 security to be bypassed and effectively broken in many situations. The only remedy as of late 2011 is to turn off Wi-Fi Protected Setup, which is not always possible.

Virtual Private Networks are often used to secure Wi-Fi.

Data Security Risks

The older wireless encryption-standard, Wired Equivalent Privacy (WEP), has been shown to be easily breakable even when correctly configured. Wi-Fi Protected Access (WPA and WPA2) encryption, which became available in devices in 2003, aimed to solve this problem. Wi-Fi access points typically default to an encryption-free (*open*) mode. Novice users benefit from a zero-configuration device that works out-of-the-box, but this default does not enable any wireless security, providing open wireless access to a LAN. To turn security on requires the user to configure the device, usually via a software graphical user interface (GUI). On unencrypted Wi-Fi networks connecting devices can monitor and record data (including personal information). Such networks can only be secured by using other means of protection, such as a VPN or secure Hypertext Transfer Protocol over Transport Layer Security (HTTPS).

Wi-Fi Protected Access encryption (WPA2) is considered secure, provided a strong passphrase is used. In 2018, WPA3 was announced as a replacement for WPA2, increasing security; it rolled out on June 26.

Piggybacking

Piggybacking refers to access to a wireless Internet connection by bringing one's own computer within the range of another's wireless connection, and using that service without the subscriber's explicit permission or knowledge.

During the early popular adoption of 802.11, providing open access points for anyone within range to use was encouraged to cultivate wireless community networks, particularly since people on average use only a fraction of their downstream bandwidth at any given time.

Recreational logging and mapping of other people's access points has become known as wardriving. Indeed, many access points are intentionally installed without security turned on so that they can be used as a free service. Providing access to one's Internet connection in this fashion may breach the Terms of Service or contract with the ISP. These activities do not result in sanctions in most jurisdictions; however, legislation and case law differ considerably across the world. A proposal to leave graffiti describing available services was called warchalking. A Florida court case determined that owner laziness was not to be a valid excuse.

Piggybacking often occurs unintentionally – a technically unfamiliar user might not change the default "unsecured" settings to their access point and operating systems can be configured to connect automatically to any available wireless network. A user who happens to start up a laptop in the vicinity of an access point may find the computer has joined the network without any visible indication. Moreover, a user intending to join one network may instead end up on another one if the latter has a stronger signal. In combination with automatic discovery of other network resources this could possibly lead wireless users to send sensitive data to the wrong middle-man when seeking a destination. For example, a user could inadvertently use an unsecure network to log into a website, thereby making the login credentials available to anyone listening, if the website uses an unsecure protocol such as plain HTTP without TLS (HTTPS).

An unauthorized user can obtain security information (factory preset passphrase and/or Wi-Fi Protected Setup PIN) from a label on a wireless access point can use this information (or connect by the Wi-Fi Protected Setup pushbutton method) to commit unauthorized and/or unlawful activities.

Infrared Communication

Infrared (IR) is a wireless mobile technology used for device communication over short ranges. IR communication has major limitations because it requires line-of-sight, has a short transmission range and is unable to penetrate walls. IR transceivers are quite cheap and serve as short-range communication solutions.

Because of IR's limitations, communication interception is difficult. In fact, Infrared Data Association (IrDA) device communication is usually exchanged on a one-to-one basis. Thus, data transmitted between IrDA devices is normally unencrypted.

IR-enabled devices are known as IrDA devices because they conform to standards set by the Infrared Data Association (IrDA). IR light-emitting diodes (LED) are used to transmit IR signals, which pass through a lens and focus into a beam of IR data. The beam source is rapidly switched on and off for data encoding.

The IR beam data is received by an IrDA device equipped with a silicon photodiode. This receiver converts the IR beam into an electric current for processing. Because IR transitions more slowly from ambient light than from a rapidly pulsating IrDA signal, the silicon photodiode can filter out the IrDA signal from ambient IR.

IrDA transmitters and receivers are classified as directed and non-directed. A transmitter or receiver that uses a focused and narrow beam is directed, whereas a transmitter or receiver that uses an omnidirectional radiation pattern is non-directed.

Infrared band of the electromagnet corresponds to 430THz to 300GHz and a wavelength of **980nm**. The propagation of light waves in this band can be used for a communication system (for transmission and reception) of data. This communication can be between two portable devices or between a portable device and a fixed device.

There are two types of Infrared communication

- Point to Point: It requires a line of sight between the transmitter and a receiver. In other words the transmitter and the receiver should be pointed to each other and there shouldn't be any obstacles between them. Example is the remote control communication.

- Diffuse Point: It doesn't require any line of sight and the link between the transmitter and the receiver is maintained by reflecting or bouncing of the transmitted signal by surfaces like ceilings, roof, etc. Example is the wireless LAN communication system

Advantages of IR communication:

- Security: Infrared communication has high directionality and can identify the source as different sources emit radiation of different frequencies and thus the risk of information being diffused is eliminated.

- Safety: Infrared radiation is not harmful to human beings. Hence infrared communication can be used at any place.

- High Speed data Communication: The data rate of Infrared communication is about 1Gbps and can be used for sending information like video signal.

IR Communication Basics:

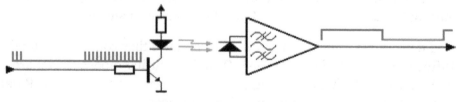

IR communication principle

IR Transmission

The transmitter of an IR LED inside its circuit, which emits infrared light for every electric pulse given to it. This pulse is generated as a button on the remote is pressed, thus completing the circuit, providing bias to the LED.

The LED on being biased emits light of the wavelength of 940nm as a series of pulses, corresponding to the button pressed. However since along with the IR LED many other sources of infrared light such as us human beings, light bulbs, sun, etc, the transmitted information can be interfered. A solution to this problem is by modulation. The transmitted signal is modulated using a carrier frequency of 38 KHz (or any other frequency between 36 to 46 KHz). The IR LED is made to oscillate at this frequency for the time duration of the pulse. The information or the light signals are pulse width modulated and are contained in the 38 KHz frequency.

IR Reception

The receiver consists of a photodetector which develops an output electrical signal as light is incident on it. The output of the detector is filtered using a narrow band filter that discards all the frequencies below or above the carrier frequency (38 KHz in this case). The filtered output is then given to the suitable device like a Microcontroller or a Microprocessor which controls devices like a PC or a Robot. The output from the filters can also be connected to the Oscilloscope to read the pulses.

Parts of IR Communication System:

IR Transmittor- IR Sensor

The sensors could be utilized as a part of measuring the radiation temperature without any contact. For different radiation temperature ranges various filters are available. An infrared (IR) sensor is an electronic device that radiates or locates infrared radiation to sense some part of its surroundings. They are undetectable to human eyes.

An infrared sensor could be considered a Polaroid that briefly recalls how an area's infrared radiation shows up. It is very regular for an infrared sensor to be coordinated into movement indicators like those utilized as a feature of private or business security systems. An IR sensor is basically has two terminals positive and negative. These sensors are undetectable to human eyes. They can measure the heat of an object and also identify movement. The region wavelength roughly from 0.75μm to 1000 μm is the IR region. The wavelength region of 0.75μm to 3 μm is called close infrared, the region from 3 μm to 6 μm is called mid infrared and the region higher than 6 μm is called far infrared. IR sensors emits at a frequency of 38 KHz.

Features of IR Sensor:

- Input voltage: 5VDC

- Sensing Range: 5cm

- Output signal: analog voltage

- Emitting element: Infrared LED

Example Interfacing Circuit of IR Diode and Photodiode

IR sensors mostly used in radiation thermometer, gas analyzers, industrial applications, IR imaging devices, tracking, and human body detection, communication and health hazards

An IR diode is connected through a resistance to the dc supply. A photo diode is connected in reverse biased condition through a potential divider of a 10k variable resistance and 1k in series to the base of the transistor. While the IR rays fall on the reverse biased photo diode it conducts that causes a voltage at the base of the transistor.

The transistor then works like a switch while the collector goes to ground. Once the IR rays are obstructed the driving voltage is not available to the transistor thus its collector goes high. This low to high logic can be used for the microcontroller input for any action as per the program.

IR Receiver/TSOP Sensor – Features & Specifications

TSOP is the standard IR remote control receiver series, supporting all major transmission codes. This is capable of receiving infrared radiation modulated at 38 kHz. IR sensors we have seen up to now working just for little short distance up to 6 cm. TSOP is sensitive to a specific frequency so its range is better contrast with ordinary photo diode. We can alter it up to 15 cm.

TSOP acts like as a receiver. It has three pins GND, Vs and OUT. GND is connected to common ground, Vs is connected to +5volts and OUT is connected to output pin. TSOP sensor has an inbuilt control circuit for amplifying the coded pulses from the IR transmitter. These are commonly used in TV remote receivers. As I said above TSOP sensors sense only a particular frequency.

Features:

- The preamplifier and photo detector both are in single package

- Internal filter for PCM frequency

- Improved shielding against electrical field disturbance

- TTL and CMOS compatibility

- Output active low

- Low power consumption

- High immunity against ambient light

- Continuous data transmission possible

Specifications:

- Supply Voltage is −0.3-6.0 V

- Supply Current is 5 mA

- Output Voltage is −0.3-6.0 V

- Output Current is 5 mA

- Storage Temperature Range is −25-+85 °C

- Operating Temperature Range is −25-+85°C

The testing of TSOP is very simple. These are commonly used in TV remote receivers. TSOP consists of a PIN diode and pre-amplifier internally. Connect TSOP sensor as shown in circuit. A LED is connected through a resistance from the supply to output.

And then when we press the button of T.V. Remote control in front of the TSOP sensor, if LED starts blinking then our TSOP sensor and its connection is correct. The point when the output of TSOP is low i.e. at the time it appropriates IR signal from a source, with a centre frequency of 38 kHz, its output goes low.

TSOP sensor is used in our daily use TV, VCD, music system's remote control. Where IR rays are transmitted by pushing a button on remote which are received by TSOP receiver inside the equipment.

References

- Kurawar, Arwa; Koul, Ayushi; Patil, Viki Tukaram (August 2014). "Survey of Bluetooth and Applications". International Journal of Advanced Research in Computer Engineering & Technology. 3: 2832–2837. ISSN 2278-1323

- Wolter Lemstra; Vic Hayes; John Groenewegen (2010). The innovation journey of Wi-Fi: the road to global success. Cambridge University Press. p. 121. ISBN 978-0-521-19971-1

- "Rise in executions for mobile use". ITV News. June 15, 2007. Archived from the original on August 17, 2007. Retrieved June 23, 2007

- Marziah Karch (1 September 2010). Android for Work: Productivity for Professionals. Apress. ISBN 978-1-4302-3000-7. Archived from the original on 17 February 2013. Retrieved 11 November 2012

- Laberge-Nadeau, Claire (September 2003). "Wireless telephones and the risk of road crashes". Accident Analysis & Prevention. 35 (5): 649–660. doi:10.1016/S0001-4575(02)00043-X

- The Committee Office, House of Commons. "Supplementary memorandum submitted by Gregory Smith". Publications.parliament.uk. Retrieved 2011-07-11

- Muller, Scott. Upgrading and Reparing PCs (22nd ed.). Pearson Education. p. 898. ISBN 978-0-7897-5610-7

- "Reaching the Mobile Respondent: Determinants of High-Level Mobile Phone Use Among a High-Coverage Group" (PDF). Social Science Computer Review. doi:10.1177/0894439309353099

- Lean, Geoffrey; Shawcross, Harriet (April 15, 2007). "Are mobile phones wiping out our bees?". The Independent. UK. Archived from the original on July 6, 2008. Retrieved May 12, 2010

Permissions

All chapters in this book are published with permission under the Creative Commons Attribution Share Alike License or equivalent. Every chapter published in this book has been scrutinized by our experts. Their significance has been extensively debated. The topics covered herein carry significant information for a comprehensive understanding. They may even be implemented as practical applications or may be referred to as a beginning point for further studies.

We would like to thank the editorial team for lending their expertise to make the book truly unique. They have played a crucial role in the development of this book. Without their invaluable contributions this book wouldn't have been possible. They have made vital efforts to compile up to date information on the varied aspects of this subject to make this book a valuable addition to the collection of many professionals and students.

This book was conceptualized with the vision of imparting up-to-date and integrated information in this field. To ensure the same, a matchless editorial board was set up. Every individual on the board went through rigorous rounds of assessment to prove their worth. After which they invested a large part of their time researching and compiling the most relevant data for our readers.

The editorial board has been involved in producing this book since its inception. They have spent rigorous hours researching and exploring the diverse topics which have resulted in the successful publishing of this book. They have passed on their knowledge of decades through this book. To expedite this challenging task, the publisher supported the team at every step. A small team of assistant editors was also appointed to further simplify the editing procedure and attain best results for the readers.

Apart from the editorial board, the designing team has also invested a significant amount of their time in understanding the subject and creating the most relevant covers. They scrutinized every image to scout for the most suitable representation of the subject and create an appropriate cover for the book.

The publishing team has been an ardent support to the editorial, designing and production team. Their endless efforts to recruit the best for this project, has resulted in the accomplishment of this book. They are a veteran in the field of academics and their pool of knowledge is as vast as their experience in printing. Their expertise and guidance has proved useful at every step. Their uncompromising quality standards have made this book an exceptional effort. Their encouragement from time to time has been an inspiration for everyone.

The publisher and the editorial board hope that this book will prove to be a valuable piece of knowledge for students, practitioners and scholars across the globe.

Index